ACTA MECHANICA

SUPPLEMENTUM 3

D. E. BESKOS AND F. ZIEGLER (EDS.)

Advances in Dynamic Systems and Stability

Festschrift for Bruno A. Boley

Springer-Verlag Wien GmbH

Prof. D. E. Beskos

Department of Civil Engineering, School of Engineering, University of Patras,
Patras, Greece

Prof. F. Ziegler

Institut für Allgemeine Mechanik, Technische Universität Wien,
Wien, Austria

Printed on acid-free paper

With 92 Figures

ISSN 0939-7906

ISBN 978-3-211-82368-2   ISBN 978-3-7091-9223-8 (eBook)
DOI 10.1007/978-3-7091-9223-8

Preface

On December 2 – 5, 1991, a Symposium on Thermal Stresses, Dynamics and Stability honoring Professor Bruno A. Boley on the occasion of his 65th birthday was held in Atlanta, Georgia during the Winter Annual Meeting of the American Society of Mechanical Engineers. The papers presented during the Symposium by some of Professor Boley's former students and colleagues cover those areas of applied mechanics where most of his contributions have been made over the years. These papers have been written in tribute to Professor Boley's distinguished scientific career and out of genuine affection and respect for him.

The present volume consists of those Symposium papers that belong to the areas of Dynamics and Stability and constitute recent advances in the field. A special issue of the Journal of Thermal Stresses has been reserved for publication of the Symposium papers on Thermal Stresses, under the editorship of Professor R. B. Hetnarski. The present volume begins with a biographical sketch and bibliography of Professor Boley, along with a list of his doctoral students.

Thirteen papers on dynamics and stability follow. The first four papers deal with wave propagation and vibration studies in solids and structures. The next two papers study wave propagation in fluids, while the seventh paper is concerned with the dynamic response of random media. Two papers dealing with structural vibrations exhibiting instability and one dealing with dynamic buckling delamination are presented next. The last three papers are concerned with instability in solids and structures.

On behalf of former students and colleagues who have had the privilege and pleasure of working with Professor Boley over the years, we express our sincere thanks for inspiring education and scholarship, as well as leadership in enhancing applied mechanics. We hope that this commemorative volume will serve as a similar inspiration to a wider audience of readers of applied mechanics.

Finally we would like to express our thanks to Professor R. B. Hetnarski, Chairman of the Symposium Commitee, to Professors F. Di Maggio, L. B. Freund, L. M. Keer, G. Maier, R. Rand and J. H. Weiner, members of the Symposium Committee, to all the authors of this volume and to Springer-Verlag, Wien, for their help and cooperation.

D. Beskos

June 1992   **F. Ziegler**

Contents

Acta Mechanica (1992) [Suppl] 3: 1—6

Bruno A. Boley: Biography

Bruno A. Boley was born in Trieste, Italy in 1924. He received the B. S. degree in Civil Engineering from the College of the City of New York in 1943, the Sc. D. degree in Aeronautical Engineering from the Polytechnic Institute of Brooklyn in 1946, and an honorary Sc. D. degree from the College of the City of New York in 1982. He served as an Assistant Professor of Aeronautical Engineering at the Polytechnic Institute of Brooklyn from 1943 to 1948. He worked at the Goodyear Aircraft Corporation from 1948 to 1950, returning then to academic life, first as an Associate Professor at the Ohio State University in the Department of Aeronautical Engineering and then, from 1952, as Professor of Civil Engineering at Columbia University.

From 1968 to 1972 he was J. P. Ripley Professor of Engineering and Chairman of the Department of Theoretical and Applied Mechanics at Cornell University. He joined the Technological Institute, Northwestern University, in January 1973 serving as Dean until 1986, and then continuing as Dean Emeritus and Walter P. Murphy Professor of Engineering. Since September 1987 he has returned to a Professorship at Columbia University. He was Visiting Professor at the Technical University of Milan (1964—65) and the Imperial College of Science and Technology, University of London (1972).

   B. A. Boley is a member of the National Academy of Engineering, and past president of the American Academy of Mechanics and of the Society of Engineering Science. He has served on the Boards of Governors of the American Society of Mechanical Engineers and Argonne National Laboratory. He is an Honorary Member of the American Society of Mechanical Engineers, and a Fellow of the American Academy of Mechanics, the American Institute of Aeronautics and Astronautics, the Society of Engineering Science, and the American Association for the Advancement of Science. He is also the founder of the Association of Chairmen of Departments of Mechanics and past Chairman of the U.S. National Committee on Theoretical and Applied Mechanics. He is Advisor-General of the International Association of Structural Mechanics in Reactor Technology (IASMiRT), has served as Secretary of the International Union of Theoretical and Applied Mechanics (IUTAM) Congress Committee. He is editor-in-chief of "Mechanics Research Communications", a technical journal published by Pergamon Press in cooperation with the International Center for Mechanical Sciences, and a member of the editorial board of twelve technical and professional journals.

He has published four books and over 100 papers in the fields of structural dynamics, elastic stability, applied mathematics, thermal stresses, heat conduction in solids, and in problems of change of phase. One of his books "Theory of Thermal Stresses", co-authored with J. H. Weiner, is listed in "Contemporary Classics in Engineering and Applied Science", ISI Press, Philadelphia, 1986, p. 316. He chaired the National Academy of Engineering's Task Force on Engineering Education in 1979 – 80. He has been the recipient of NSF and NATO Fellowships, the Townsend Harris Medal of the Alumni Association of the City of New York, the Th. von Karman Medal of the American Society of Civil Engineers and the Worcester Reed Warner Medal of the American Society of Mechanical Engineers for outstanding contribution to the permanent literature of Engineering.

List of publications of Bruno A. Boley

Books

[1] Boley, B. A., Weiner, J. H.: Theory of thermal stresses. New York: Wiley 1960.
[2] Freudenthal, A. M., Boley, B. A., Liebowitz, H. (eds.): High temperature structures and materials. Proceedings of 3rd symposium on naval structural mechanics. New York: Pergamon Press 1964.
[3] Boley, B. A. (ed.): Thermoinelasticity. Berlin, Heidelberg, New York: Springer 1970.
[4] Boley, B. A. (ed.): Crossfire in professional education: students, the professions and society. New York: Pergamon Press 1977.

Articles

[1] Hoff, N. J., Boley, B. A.: The shearing rigidity of curved panels under compression. N.A.C.A. technical note 1090. Washington D.C.: N.A.C.A., August 1946.
[2] Hoff, N. J., Boley, B. A. et al.: Stresses in and general instability of monocoque cylinders with cutouts. Part I: N.A.C.A. technical note 1013, June 1946; Part II: N.A.C.A. technical note 1014, June 1946; Part III: N.A.C.A. technical note 1263, May 1947; Part IV: N.A.C.A. technical note 1264, February 1948; Part VI: N.A.C.A. technical note 1436, March 1948; Part VII: N.A.C.A. technical note 1962, October 1949; Part VIII: N.A.C.A. technical note 1963, October 1949. Washington D.C.: N.A.C.A. 1946 – 1949.
[3] Boley, B. A.: Numerical methods for the calculation of elastic instability. J. Aeronaut. Sci. **14**, 337 – 348 (1947).
[4] Boley, B. A., Lieber, P.: An analysis of two-dimensional viscous flow with shockwaves. Polytechnic Institute of Brooklyn Aeronautical Laboratory Report No 137. Brooklyn, New York: Polytechnic Institute of Brooklyn, October 1948.
[5] Hoff, N. J., Boley, B. A., et al.: The inward bulge type buckling of monocoque cylinders. Part IV: N.A.C.A. technical note 1499, Part V: N.A.C.A. technical note 1505. Washington D.C.: N.A.C.A. September 1948.
[6] Hoff, N. J., Boley, B. A., Coan, J. M.: The development of a technique for testing stiff panels in edgewise compression. Proceedings of the Society for Experimental Stress Analysis **5**, 14 – 24 (1948).

[7] Boley, B. A.: Concentrated loads effects in reinforced monocoque cylinders-consideration of the eccentricity and the shearing and extensional deformations of the rings, In: Hans Reissner anniversary volume (edited by Department of Aeronautical Engineering and Applied Mechanics of Polytechnic Institute of Brooklyn) pp. 313 – 320. Ann Arbor, Michigan: J. W. Edwards 1949.

[8] Hoff, N. J., Salerno, V. L., Boley, B. A.: Shear stress concentration and moment reduction factors for reinforced monocoque cylinders subjected to concentrated radial loads. J. Aeronaut. Sci. **16**, 277 – 288 (1949).

[9] Boley, B. A.: The shearing rigidity of buckled sheet panels. J. Aeronaut. Sci. **17**, 356 – 363 (1950).

[10] Boley, B. A., Moore, R. H.: A note on the calculation of deflections of indeterminate structures. J. Aeronaut. Sci. **17**, 526 – 527 (1950).

[11] Boley, B. A., Kempner, J., Meyers, J.: A numerical approach to the instability problem of monocoque cylinder. N.A.C.A. technical note 2354, Washington D.C.: N.A.C.A. April 1951.

[12] Boley, B. A., Moore, R. H.: The calculation of deflections of indeterminate structures. J. Aeronaut. Sci. **18**, 285 (1951).

[13] Hoff, N. J., Boley, B. A., Nardo, S. V., Kaufman, S.: The buckling of rigid-jointed plane trusses. Trans. Am. Soc. Civ. Eng. **116**, 958 – 998 (1951).

[14] Boley, B. A., Zimmoch, V. P.: Lateral buckling of non-uniform beams. J. Aeronaut. Sci. **19**, 567 – 568 (1952).

[15] Boley, B. A.: Graphical-numerical solution of problems of Saint Venant torsion and bending. J. Appl. Mech. **20**, 321 – 326 (1953).

[16] Boley, B. A.: An energy theory of transverse impact on beams. Proceedings of the 8th International Congress on Theoretical and Applied Mechanics, Istanbul, Turkey, 1953.

[17] Boley, B. A.: An approximate theory of lateral impact on beams. J. Appl. Mech. **22**, 69 – 76 (1955).

[18] Boley, B. A.: The application of Saint Venant's principle in dynamical problems. J. Appl. Mech. **22**, 204 – 206 (1955).

[19] Boley, B. A., Chao, C. C.: Some solutions of the Timoshenko beam equations. J. Appl. Mech. **22**, 579 – 586 (1955).

[20] Boley, B. A.: The determination of temperature, stresses and deflections in two-dimensional thermoelastic problems. J. Aeronaut. Sci. **23**, 67 – 75 (1956).

[21] Boley, B. A.: Thermally induced vibrations of beams. J. Aeronaut. Sci. **23**, 179 – 181 (1956).

[22] Boley, B. A.: A method for the construction of Green's functions. Quarter. Appl. Math. **14**, 249 – 257 (1956).

[23] Boley, B. A., Tolins, I. S.: On the stresses and deflections of rectangular beams. J. Appl. Mech. **23**, 339 – 342 (1956).

[24] Boley, B. A.: On thin-ring analysis. J. Aeronaut. Sci. **23**, 802 – 803 (1956).

[25] Boley, B. A., Chao, C. C.: Impact on pin-jointed trusses. Trans. Am. Soc. Civ. Eng. **122**, 39 – 61 (1957).

[26] Boley, B. A.: The calculation of thermoelastic beam deflections by the principle of virtual work. J. Aeronaut. Sci. **24**, 139 – 141 (1957).

[27] Barber, A. D., Weiner, J. H., Boley, B. A.: An analysis of the effect of thermal contact resistance in a sheet-stringer structure. J. Aeronaut. Sci. **24**, 232 – 233 (1957).

[28] Weiner, J. H., Boley, B. A.: Basic concepts of the thermodynamics of continuous media. Technical report No 57 – 288 Wright Aeronautics Development Command (WADC), Dayton, Ohio: Wright-Patterson Air Force Base 1957.

[29] Boley, B. A.: On the use of sine transforms in Timoshenko beams impact problems. J. Appl. Mech. **24**, 152 – 153 (1957).

[30] Boley, B. A., Barber, A. D.: Dynamic response of beams and plates to rapid heating. J. Appl. Mech. **24**, 413 – 416 (1957).

[31] Boley, B. A.: A method for the construction of fundamental solutions in elasticity theory. J. Math. Phys. **36**, 261 – 267 (1957).

[32] Boley, B. A.: A method for the numerical evaluation of certain infinite integrals. Math. Tables and Other Aids to Computation **9**, 261 – 268 (1957).

[33] Boley, B. A., Barrekette, E. S.: Thermal stresses in curved beams. J. Aerosp. Sci. **25**, 627 – 631 (1958).

[34] Boley, B. A., Chao, C. C.: An approximate analysis of Timoshenko beams under dynamic loads. J. Appl. Mech. **25**, 31 – 36 (1958).

[35] Boley, B. A.: Some observations on Saint Venant's principle. In: Proceedings of the 3rd U.S. National Congress of Applied Mechanics, pp. 259 – 264, New York: American Society of Mechanical Engineers 1958.

[36] Boley, B. A.: Critical aspects of air frame design and production. In: Inspection for disarmament (Melman, S., ed.) pp. 139—146, New York: Columbia University Press 1958.

[37] Boley, B. A., Friedman, M. B.: On the viscous flow around the leading edge of a flat plate. J. Aerosp. Sci. **26**, 453—454 (1959).

[38] Boley, B. A.: Thermal stresses, In: Structural mechanics. Proceedings of 1st symposium on naval structural mechanics (Goodier, J. N., Hoff, N. J., eds.) pp. 378—406. New York: Pergamon Press 1960.

[39] Boley, B. A.: Upper bounds and Saint Venant's principle in transient heat conduction. Quarter. Appl. Math. **18**, 205—207 (1960).

[40] Boley, B. A.: On a dynamical Saint Venant's principle. J. Appl. Mech. **27**, 74—78 (1960).

[41] Boley, B. A.: A method of heat conduction analysis of melting and solidification problems. J. Math. Phys. **40**, 300—313 (1961).

[42] Boley, B. A.: Discontinuities in integral-transform solutions. Quarter. Appl. Math. **19**, 273—284 (1962).

[43] Boley, B. A.: Transient coupled thermoelastic boundary-value problems in the half-space. J. Appl. Mech. **29**, 637—646 (1962).

[44] Boley, B. A.: Some bounds for steady-state temperature distributions, In: Proceedings of 4th U.S. national congress of applied mechanics, pp. 1197—1204, New York: American Society of Mechanical Engineers 1962.

[45] Weiner, J. H., Boley, B. A.: Elasto-plastic thermal stresses in a solidifying slab. J. Mech. Phys. Solids **11**, 145—154 (1963).

[46] Boley, B. A.: A post-doctoral preceptorial program. J. Eng. Educ. **54**, 129—130 (1963).

[47] Boley, B. A.: Upper and lower bounds for the solution of a melting problem. Quarter. Appl. Math. **21**, 1—11 (1963).

[48] Boley, B. A.: On the accuracy of the Bernoulli-Euler theory for beams of variable section. J. Appl. Mech. **30**, 373—378 (1963).

[49] Boley, B. A.: The analysis of problems of heat conduction and melting, In: High temperature structures and materials. Proceedings of 3rd symposium on naval structural mechanics (Freudenthal, A. M., Boley, B. A., Liebowitz, H., eds.) pp. 260—315. New York: Pergamon Press 1964.

[50] Boley, B. A.: Upper and lower bounds in problems of melting or solidifying slabs. Quarter. J. Mech. Appl. Math. **17**, 253—269 (1964).

[51] Boley, B. A.: Estimate of errors in approximate temperature and thermal stress calculation, In: Proceedings of 11th International Congress of Theoretical and Applied Mechanics (Görtler, H., ed.) pp. 586—596. Berlin, Göttingen, Heidelberg, New York: Springer 1964.

[52] Sikarskie, D. L., Boley, B. A.: The solution of a class of two-dimensional melting and solidification problems. Int. J. Solids Struct. **1**, 207—234 (1965).

[53] Boley, B. A., Weiner, J. H.: Thermal stresses. In: Metals engineering design (Horger, O. J., ed.) pp. 106—122. New York: McGraw-Hill 1965.

[54] Wu, T. S., Boley, B. A.: Bounds in melting problems with arbitrary rates of liquid removal. SIAM J. on Applied Mathematics **14**, 306—323 (1966).

[55] Boley, B. A.: Bounds on the maximum thermoelastic stress and deflection in a beam or plate. J. Appl. Mech. **33**, 881—887 (1966).

[56] Spitzer, L., Jr., Boley, B. A.: Thermal deformations in a satellite telescope mirror. J. Opt. Soc. Am. **57**, 901—913 (1967).

[57] Boley, B. A., Hetnarski, R. B.: Propagation of discontinuities in coupled thermoelastic problems. J. Appl. Mech. **35**, 489—494 (1968).

[58] Boley, B. A.: A general starting solution for melting or solidifying slabs. Int. J. Eng. Sci. **6**, 89—111 (1968).

[59] Boley, B. A.: Thermoelasticity. In: Advisory Group for Aeronautical Research and Development (AGARD) manual on aeroelasticity, Paris: North Atlantic Treaty Organization (N.A.T.O.), 1968.

[60] Boley, B. A., Yagoda, H. P.: The starting solution for two-dimensional heat conduction problems with change of phase. Quarter. Appl. Math. **27**, 223—246 (1969).

[61] Boley, B. A., Testa, R. B.: Thermal stresses in composite beams. Int. J. Solids Struct. **5**, 1153—1169 (1969).

[62] Boley, B. A.: On thermal stresses and deflections in thin rings. Int. J. Mech. Sci. **11**, 781—789 (1969).

[63] Patel, P. D., Boley, B. A.: Solidification problems with space and time varying boundary conditions and imperfect mold contact. Int. J. Eng. Sci. **7**, 1041—1066 (1969).

[64] Testa, R. B., Boley, B. A.: Basic thermoelastic problems in fiber-reinforced materials. In: Mechanics of composite materials. Proceedings of 5th symposium on naval structural mechanics (Wendt, F. W., Liebowitz, H., Perrone, N., eds.) pp. 361—385, New York: Pergamon Press 1970.

[65] Boley, B. A.: Uniqueness in a melting slab with space- and time-dependent heating. Quarter. of Appl. Math. **27**, 481−487 (1970).

[66] Lederman, J. M., Boley, B. A.: Axisymmetric melting or solidification of circular cylinders. Int. J. Heat Mass Transfer **13**, 413−427 (1970).

[67] Grimado, P. B., Boley, B. A.: A numerical solution for the symmetric melting of spheres. Int. J. Numer. Methods Eng. **2**, 175−188 (1970).

[68] Friedman, E., Boley, B. A.: Stresses and deformations in melting plates. J. Spacecraft. Rockets **7**, 324−333 (1970).

[69] Boley, B. A., Lederman, J. M., Grimado, P. B.: Radially symmetric melting of cylinders and spheres, In: Proceedings of 4th international heat transfer conference, Versailles, France. paper Cu 2.2., 1−11. Amsterdam: Elsevier 1970.

[70] Yagoda, H. P., Boley, B. A.: Starting solutions for melting of a slab under plane or axisymmetric hot spots. Quarter. J. Mech. Appl. Math. **23**, 225−246 (1970).

[71] Boley, B. A.: Instability of bars with stress-dependent properties. In: IUTAM symposium on instability of continuous systems (Leipholz, H., ed.) pp. 90−95. Berlin, Heidelberg, New York: Springer 1971.

[72] Boley, B. A.: Bounds for the torsional rigidity of heated beams. American Institute of Aeronautics and Astronautics J. **9**, 524−525 (1971).

[73] Boley, B. A.: Preface to Aerostructures-selected papers of Nicholas J. Hoff (Testa, R. B., ed.) New York: Pergamon Press 1971.

[74] Boley, B. A., Yagoda, H. P.: The three-dimensional starting solution for a melting slab. Proc. R. Soc. London **A 323**, 89−110 (1971).

[75] Boley, B. A.: Temperature and deformations in rods and plates melting under internal heat generation, In: Proceedings of 1st international conference on structural mechanics in reactor technology (Jaeger, T. A., ed.) Vol. 5, paper L 2/3. Amsterdam: North Holland 1971.

[76] Boley, B. A.: Biographies of A. Castigliano, L. Menabrea, G. Poleni and I. Porro, In: Dictionary of scientific biography (Gillispie, C. C., ed.) New York: Charles Scribner's Sons 1971.

[77] Boley, B. A.: On a melting problem with temperature-dependent properties, In: Trends in elasticity and thermoelasticity. W. Nowacki anniversary volume (Czarnota-Bojarski, R. E., Sokolowski, M., Zorski, H., eds.) pp. 22−28. Groningen, The Netherlands: Wolters-Noordhoff 1971.

[78] Boley, B. A.: Approximate analyses of thermally induced vibrations of beams and plates. J. Appl. Mech. **39**, 212−216 (1972).

[79] Boley, B. A.: Survey of recent developments in the fields of heat conduction in solids and thermoelasticity. Nuc. Eng. Des. **18**, 377−399 (1972).

[80] Boley, B. A.: On thermal stresses in beams: some limitations of the elementary theory. Int. J. Solids Struct. **8**, 571−579 (1972).

[81] Testa, R. B., Bennett, B., Boley, B. A.: Bounds on thermal stresses in composite beams of arbitrary cross-section. Int. J. Solids Struct. **8**, 907−912 (1972).

[82] Boley, B. A., Lee, Y. F.: A problem in heat conduction of melting plates. Lett. Appl. Engineering Sciences **1**, 25−32 (1973).

[83] Boley, B. A.: On the use of superposition in the approximate solution of heat conduction problems. Int. J. Heat Mass Transfer **16**, 2035−2041 (1973).

[84] Boley, B. A.: Applied thermoelasticity. Dev. Mech. **7**, 3−16 (1973).

[85] Dökmeci, M. C., Boley, B. A.: Vibration analysis of a rectangular plate. J. Franklin Inst. **296**, 305−321 (1973).

[86] Boley, B. A.: Methods of approximate heat conduction analysis, In: Proceedings of 2nd international conference on structural mechanics in reactor technology (Jaeger, T. A., ed.) Vol. 5, paper L 1/2. Amsterdam: North Holland 1973.

[87] Boley, B. A., Lee, Y. F.: Melting of an infinite solid with a spherical cavity. Int. J. Eng. Sci. **11**, 1277−1295 (1973).

[88] Boley, B. A.: Thermal stresses today, In: Proceedings of 7th U.S. national congress of applied mechanics (Datta, S. K., ed.) pp. 99−107. New York: American Society of Mechanical Engineers 1974.

[89] Boley, B. A.: The embedding technique in melting and solidification problems, In: Moving problems in heat flow and diffusion (Ockendon, J. R., Jr., Hodgkins, W. R., eds.) pp. 150−172. Oxford: Clarendon Press 1975.

[90] Lahoud, A., Boley, B. A.: Some considerations on the melting of reactor fuel plates and rods. Nuc. Eng. Des. **32**, 1−19 (1975).

[91] Beskos, D. E., Boley, B. A.: Use of dynamic influence coefficients in forced vibration problems with the aid of Laplace transform. Comput. Struct. **5**, 263−269 (1975).

[92] Beskos, D. E., Boley, B. A.: The Gauss-Seidel convergence criterion for elastic structural stability. Mechanics Research Communications **3**, 457−462 (1976).

[93] Tsubaki, T., Boley, B. A.: One-dimensional solidification of binary mixtures. Mechanics Research Communications **4**, 115−122 (1977).

[94] Boley, B. A., Estenssoro, L.: Improvements on approximate solutions in heat conduction. Mechanics Research Communications **4**, 271−279 (1977).

[95] Boley, B. A.: An applied overview of moving boundary problems, In: Proceedings of the symposium and workshop on moving boundary problems, Gatlinburg, Tennessee (Wilson, D. G., Solomon, A. D., eds.) pp. 205−231. New York: Academic Press 1977.

[96] Boley, B. A.: Some considerations of elastic analyses of discrete models of solids. Comput. Struct. **8**, 345−347 (1977).

[97] Boley, B. A.: Time dependent solidification of binary mixtures. Int. J. Heat Mass Transfer **21**, 821−824 (1978).

[98] Boley, B. A., Estenssoro, L.: An approximate representation of Neumann's solidification solution. Mechanics Research Communications **6**, 245−250 (1979).

[99] Beskos, D. E., Boley, B. A.: Critical damping of linear discrete dynamic systems. J. Appl. Mech. **47**, 627−630 (1980).

[100] Boley, B. A.: Education of engineers versus productivity, In: Proceedings of a national conference on improving the productivity of technical resources (Skan, L., ed.) pp. 1−7. Illinois Institute of Technology Research Institute, Chicago, Illinois, November 1980.

[101] Boley, B. A.: Thermal stresses: a survey, In: Thermal stresses in severe environments (Hasselman, D. P. H., Heller, R. A., eds.) pp. 1−11. New York: Plenum Publishing Corporation 1980.

[102] Beskos, D. E., Boley, B. A.: Critical damping in certain linear continuous dynamic systems. Int. J. Solids Struct. **17**, 575−588 (1981).

[103] Boley, B. A.: On atomic spacing in large regular cubic lattices, In: Micromechanics and inhomogeneity. T. Mura anniversary volume (Weng, G. J., Taya, M., Abe, H., eds.) pp. 83−88. Berlin, Heidelberg, New York, Tokyo: Springer 1989.

[104] Boley, B. A.: An approximate calculation of atomic spacing in a large regular lattice, In: Future trends in applied mechanics. P. S. Theocaris anniversary volume. pp. 51−60. Athens: National Technical University of Athens 1989.

[105] Boley, B. A.: Elastic moduli of atomic lattices, In: Proceedings of international conference on mechanics, physics and structure of materials. Aristotle 23 centuries celebration (Aifantis, E. C., ed.) p. 5. Thessaloniki: University of Thessaloniki 1990.

[106] Boley, B. A.: Some thoughts on the analysis of solids on the basis of atomic lattice models, In: Computational mechanics '91: theory and applications (Atluri, S. N., Beskos, D. E., Jones, R., Yagawa, G., eds.) pp. 131−136. Atlanta: International Conference of Computational Engineering Science (ICES) Publications 1991.

List of doctoral students of Bruno A. Boley

C.-C. Chao	1958	Columbia University
E. S. Barrekette	1959	Columbia University
S. J. Citron	1959	Columbia University
I. S. Tolins	1961	Columbia University
D. L. Sikarskie	1964	Columbia University
T.-S. Wu	1964	Columbia University
E. Friedman	1965	Columbia University
H. P. Yagoda	1967	Columbia University
P. B. Grimado	1968	Columbia University
J. M. Lederman	1968	Columbia University
P. D. Patel	1968	Columbia University
Y. F. Lee	1971	Cornell University
M. C. Dökmeci	1972	Cornell University
A. Lahoud	1973	Cornell University
D. E. Beskos	1973	Cornell University

Acta Mechanica (1992) [Suppl] 3: 7−21

Wave propagation in composite structures: experimental and theoretical studies

M. B. Sayir, Zurich, Switzerland

Summary. Structural waves with well-defined time functions involving relatively large wavelengths with respect to the thickness are generated at some adequately chosen part of composite structures such as filament-wound tubes or fiber-reinforced cross-ply laminates. Theoretical predictions based on asymptotic analysis of the basic three-dimensional equations of elastodynamics are compared with experimental results and allow a thorough understanding of the main physical mechanisms of structural wave propagation in composite structures.

1 Introduction

Structural waves (also called "low-frequency guided waves") in thin structures such as plates and shells are characterized by wavelengths sufficiently large with respect to the thickness of the structures in which they propagate. Inertial effects due to "breathing modes" across the thickness can then be neglected. Only membrane and flexural modes with simple distributions of stresses and displacements across the thickness are relevant. For linear elastic, isotropic and homogeneous behaviour extensive information based on both theoretical and experimental studies is readily available in the classic literature on linear wave propagation in plates and shells (see for example [1] to [3]). For composite structures such as filament-wound tubes or fiber-reinforced laminates (and sandwich-plates), due to various coupling effects, both theoretical and experimental aspects of structural wave propagation are more complex. For such structures attention has been mostly focused on propagation of ultrasonic waves with wavelengths much shorter than the thickness of the structure (see for example [4] and [5]).

In the past decade, motivated by possible applications in non-destructive quantitative evaluation procedures, we have developed at our Institute in Zurich efficient experimental techniques to study with considerable precision structural waves in composite structures. We use hereby simple but sufficiently accurate theoretical models of structural dynamic behaviour to analyze the results and to design the experiments. In this paper I will shortly review some of the results obtained in *two typical examples*. The following features are common to both cases:

(1) The waves were generated with the help of piezoelectric transducers which were excited with well-defined time-functions corresponding to *narrow-band amplitude-frequency spectra* (Fig. 1). This is an important feature increasing both the accuracy of the measured dispersion behaviour and the "comfort" of the corresponding theoretical analysis, with respect to other techniques using for example impact loading or even capacitive or inductive excitation.

(2) The above-mentioned frequency-bands were *well below the lowest resonance frequency of the transducer*. Thus possible elastic coupling between the wave generator and the responding

structure was reduced to a minimum. This also decreases considerably interferences with spurious oscillations caused by environmental factors.

(3) The generated *displacement amplitudes were of the order of fractions of μm*, so that linearity assumptions hold with extreme accuracy.

(4) The response of the structure was measured optically with the help of a self-built *heterodyne-type Laser-Doppler-Interferometer* (Figs. 2 and 3). Most of the needed information concerned *frequencies and phases* which can be measured and analyzed with greater accuracy than information relying on the analysis of amplitudes.

(5) Thanks to feature (1) a given experiment could be repeated hundreds of times in a relatively short time interval (a few seconds). Simple averaging allowed then a considerable *decrease in the noise to signal ratio*. Choosing *frequencies in the range above 3−5 kHz* (upper limits are essentially imposed by restriction (2) concerning the lowest resonant frequency of the

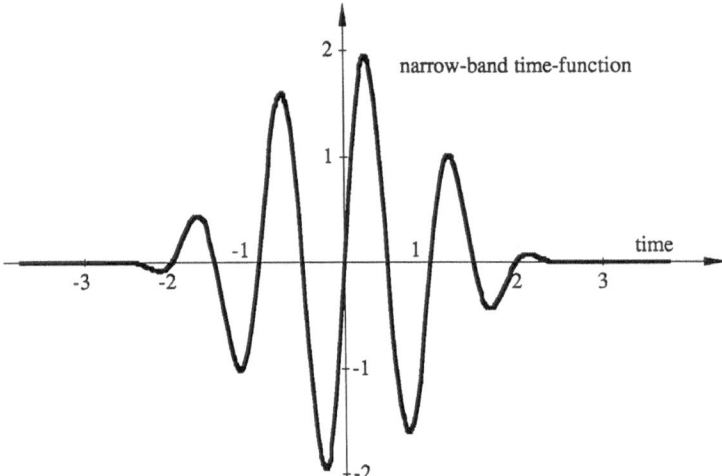

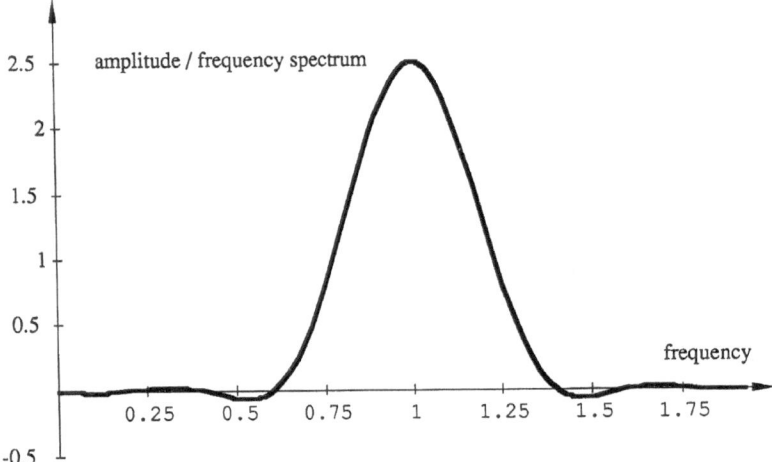

Fig. 1. Narrow-band time function and its amplitude/frequency spectrum. The time function illustrated here is $F(t) := \left\{ a \left[1 + \cos \left(2\pi \frac{f_0}{N} t \right) \right] \sin (2\pi f_0 t), \text{ for } |t| \leq \frac{N}{2f_0}, 0 \text{ otherwise} \right\}$ and corresponds to a "Hanning-window" superimposed on a sinusoid. For nondimensional representation the amplitude a and the fundamental frequency f_0 have been chosen as 1. The number of periods in the window has been set here as $N = 5$ which is the value we usually choose in our experiments. Increasing N reduces the width of the band but lengthens the duration of the pulse. This could cause problems in connection with reflections from the boundaries

transducer; we usually remained below 200 kHz) also contributed to decrease this ratio. A typical measurement is shown in Fig. 4.

(6) *Dispersion curves* relating phase velocities c to frequencies f (or to "wave-numbers" $2\pi/\Lambda$, where Λ is the wavelength $\Lambda = c/f$) were obtained by *Fourier-Analysis*, comparing the phase spectra in the relevant frequency intervals of two response functions measured at two conveniently chosen points of the structure. Hereby attention was paid to *avoid interference with possible reflections from the boundaries* of the structure.

(7) The quantitative theoretical evaluation of the experimental results was based in all three cases on careful *asymptotic analysis* (using multi-parameter perturbation procedures by proper scaling) of the basic three-dimensional equations of linear elastodynamics. This is particularly important in the case of composite structures, where not only the usual *small dimensionless geometric parameter* $\varepsilon_\Lambda := H/\Lambda$ (H: thickness of the structure, Λ: wavelength corresponding to the fundamental frequency of the narrow-band time-function) determines the structural dynamic behaviour but also a *small dimensionless material parameter* $\varepsilon_M := G/E$ (G: "small" shear modulus, E: "large" tensile modulus). Thus simplified sets of equations for dynamic structural behaviour do not have to be introduced by a priori assumptions (for example modified Kirchhoff-Love assumptions), they follow directly from the "ordering algorithm" of the asymptotic expansions (see for example [6] for the case of unidirectionally fiber-reinforced beams). *Coupling due to shear deformation* can then be detected and accounted for in the proper mathematical form following from the equations.

(8) Theoretical evaluations based on asymptotic analysis lead quite often to a better understanding and *prediction of the different physical mechanisms*. Indeed, exact solutions, when available, deliver quite often relations which are too involved to be of use in straightforward

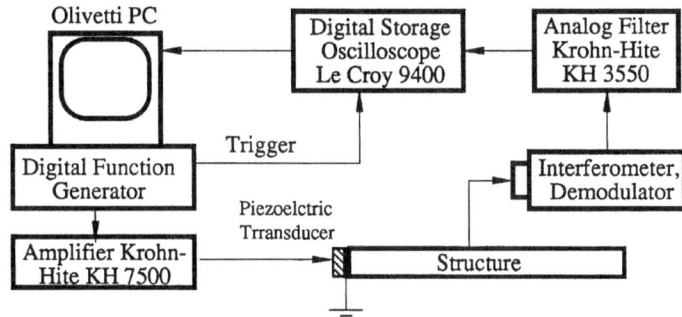

Fig. 2. A schematic description of the experimental set-up for measuring structural waves

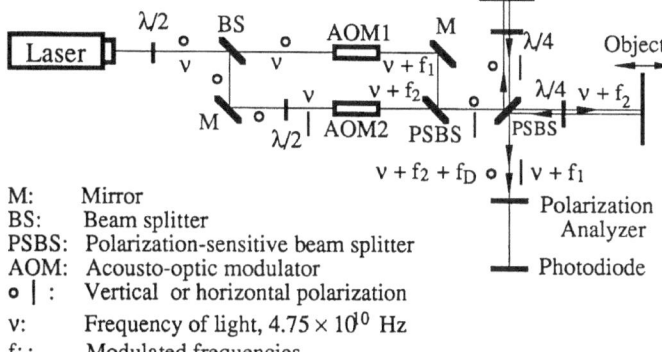

Fig. 3. Heterodyne Laser-Doppler Interferometer used in structural wave measurements

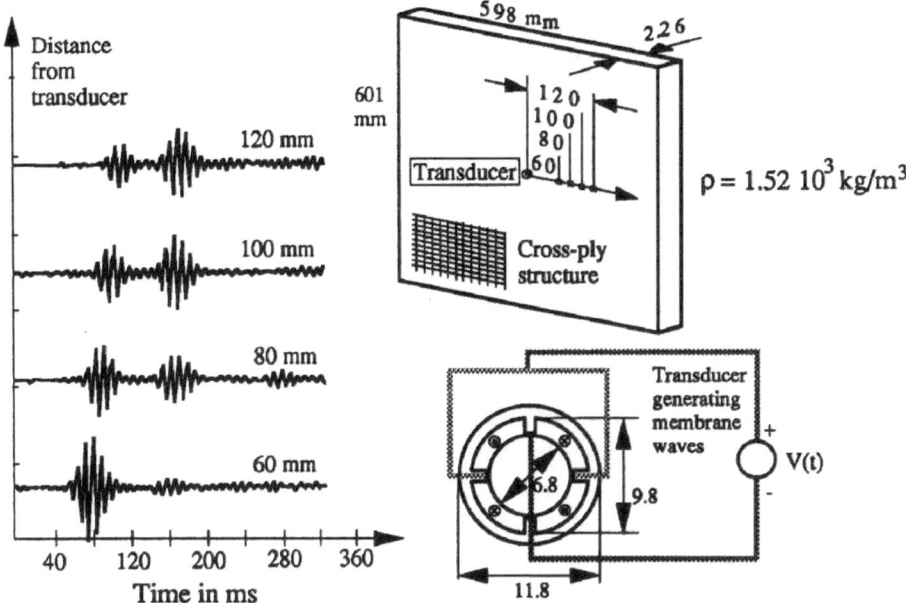

Fig. 4. A typical measurement of structural wave propagation. The illustration shown here corresponds to in-plane displacements in one of the two principal directions of a composite fiber-reinforced cross-ply laminate measured at radial distances from a ring transducer (shear transducer) producing originally displacements in its hoop direction (see Sect. 2)

physical interpretations and can only made explicit by numerical evaluation. In contrast, asymptotic analysis allows a better separation of the physical phenomena and hence leads to simpler theoretical results which lend themselves to more comfortable physical implementation. This will be illustrated in the two examples presented in Sects. 2 and 3.

2 Structural waves in filament-wound tubes

2.1 Experimental preliminaries

The experiments were carried out at our Institute by *J. Dual* as part of the work leading to his PhD thesis [7]. He used 10 carbon-fibre reinforced tubes (with an epoxy matrix) produced by filament winding on a total length of 2 m with 12 000 filaments per roving and a total number of 15 layers (Fig. 5). For each of the 5 cases of winding angles of $0°$, $22.5°$, $45°$, $67.5°$ and $82°$ (with respect to the axis) two tubes were available. While the mean radius and the wall thickness were fairly constant along the length and the hoop direction of each tube, the value of the mean radius R varied from 14.72 to 14.84 mm between the tubes and the wall thickness H from 1.5 to 1.7 mm.

The main purpose of the experiments was to measure the dispersion behaviour (wave-number or phase velocity as a function of frequency) of the tubes. The "low" frequency range corresponding to frequencies (maximum of 25 to 40 kHz depending on the winding angle of the tube) leading to wavelengths larger than the perimeter of the tube ($\Lambda > 2\pi R$) was covered by Dual with the help of resonance measurements. Indeed at such relatively low frequencies the signal of the transducer would have been too weak to produce acceptable signal to noise ratios in transient wave experiments. On the other side the detection of resonance frequencies at higher modes becomes quite problematic in the middle and higher range, since those frequencies are too close to each other to allow separation. Even the particularly efficient method of resonance-

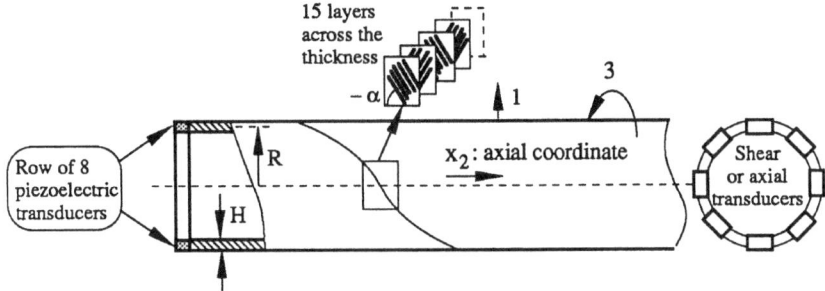

Fig. 5. Filament-wound cylindrical shell axisymmetrically loaded at one end by piezoelectric transducers in radial shear or axial stress

measurement which is implemented in our laboratory (based on fixing the frequency at 90° phase angle rather than looking for the maximum of the amplitude as is usual in more classical experiments) is of little help, if resonance frequencies of the order of 30 kHz are separated only by 100 Hz or less and thus coupling by internal damping becomes very strong.

In the middle and higher frequency range leading to wavelengths smaller than the perimeter of the tube but still sufficiently large with respect to the wall thickness (to ensure structural behaviour) Dual determined wave-numbers $2\pi/\Lambda$ as a function of frequency by generating and measuring transient waves with the help of the experimental setup illustrated in Figs. 2 and 3. He applied hereby Fourier-Analysis to narrow-band pulses (Fig. 1) with various fundamental frequencies and determined the wave numbers as a function of frequency by comparing the phase spectra of the pulses measured at two or more cross sections of the tube. The success of the measurements depends very much on the proper choice of the frequency range, of the mode of excitation, of the positioning of the interferometer etc. For these choices a thorough theoretical understanding of the expected dynamic behaviour of the structure turned out to be crucial. In the following subsection theoretical results based on asymptotic analysis of the basic three-dimensional equations of elastodynamics will be shortly reviewed. A more detailed account will be presented elsewhere [8].

2.2 Axisymmetric structural waves in orthotropic cylindrical shells

Since the wavelengths are sufficiently large with respect of the length scale of the micro-structure (thickness of the filaments and rovings) of the tube, homogeneous elastodynamic behaviour may be assumed. The following three small dimensionless parameters are introduced by proper scaling of the coordinates, the stresses and the displacements in the well-known three-dimensional basic equations of linear elastodynamics for homogeneous orthotropic constitutive behaviour and axisymmetric geometry:

$$\varepsilon_R := \frac{H}{R}, \quad \varepsilon_A := \frac{H}{\Lambda}, \quad \varepsilon_M := \frac{G_{12}}{E_2}, \tag{1.1}$$

where E_2 is the modulus of elasticity in the axial direction and G_{12} is the transverse shear modulus (across the wall thickness). The first parameter is small since the tube is supposed to be a "thin-walled" shell. The second parameter is small if in the induced narrow-band frequency spectrum only frequencies up to moderately high values are represented. In the present case frequencies up to 250 kHz (this "limit" depends on the winding angle) were allowed. The smallness of ε_A is of course a prerequisite for structural behaviour. The third parameter is small because in the structures considered here the matrix which determines decisively the value of the transverse shear stiffness is supposed to be much softer than the filaments reinforcing it. In the following we will assume

$$\varepsilon_M = O(\varepsilon_R). \tag{1.2}$$

This assumption is satisfactorily fulfilled for the tubes with $0°$, $22.5°$ and $45°$ winding angles ($\varepsilon_R \approx 0.1$ and $\varepsilon_M \approx 0.04, 0.06, 0.2$ respectively as confirmed a posteriori by the measurements leading to the values of E_2 and G_{12}). It is also applicable to the remaining two cases with larger winding angles if on replaces E_2 by the elasticity modulus E_3 (in the hoop direction) in the definition of ε_M given in (1.1). The degree of anisotropy corresponding to the assumption above has been called "*moderately strong*" in [6].

As mentioned at the end of the previous subsection, the details of the asymptotic analysis will be presented in [8]. Here we list the main results of the analysis restricted to the *first step of approximation* in the asymptotic sense. Three cases arise as long as ε_A remains small:

I) Wavelengths comparable with the perimeter of the tube:

$$2\pi R = O(\Lambda) \quad \text{or} \quad \varepsilon_A = O(\varepsilon_R).$$

The distribution of the displacement components u_1 (radial) and u_2 (axial) across the thickness of the tube wall turns out to be constant for the first step of approximation. Of course the component u_3 the hoop direction vanishes identically because of axisymmetry. Such a state of deformation is usually called "*membrane mode*". Indeed in this case the flexural stiffness of the shell wall can be neglected since it contributes only terms of order $O[(\varepsilon_A)^2]$ with respect to other terms, the shell behaves like a membrane. The following differential equations of the first asymptotic step of approximation can be shown to hold for the displacement components:

$$-u_1 - vRu_{2,2} = R^2(c_3)^{-2} (1 - ev^2) u_{1,tt},$$

$$vu_{1,2} + R\,e^{-1}u_{2,22} = R(c_3)^{-2} (1 - ev^2) u_{2,tt}. \tag{1.3}$$

Here a comma has been used to indicate partial differentiation with respect to the corresponding variable (axial coordinate x_2 or time t), the parameter v is the usual Poisson-coefficient for lateral contraction and corresponds in this case of orthotropy to v_{23} (the indices have been dropped to alleviate the formalism), the quantity c_3 is the longitudinal wave velocity in the hoop direction under uniaxial stress and is defined as

$$c_3 := \sqrt{\frac{E_3}{\varrho}}, \tag{1.4}$$

where E_3 is the modulus of elasticity in the hoop direction and ϱ is the specific mass per unit volume. The number e is the ratio of the two moduli

$$e := E_3/E_2. \tag{1.5}$$

For sinusoidal waves the dispersion relation relating phase velocities to frequencies can easily be derived from (1.3). Denote by Ω the dimensionless frequency defined as

$$\Omega := \frac{2\pi R}{c_3} f \tag{1.6}$$

(f: frequency in Hertz) and by C the dimensionless phase velocity defined as

$$C = c/c_2 \qquad (1.7)$$

where c is the current phase velocity in m/s and

$$c_2 := \sqrt{\frac{E_2}{\varrho}} \qquad (1.8)$$

is the longitudinal wave velocity in the axial direction under uniaxial states of stress. The dispersion relation derived from (1.3) is then

$$C^2 = \frac{1 - \Omega^2}{1 - \Omega^2(1 - ev^2)}. \qquad (1.9)$$

This expression is very similar to the one which can be derived for isotropic behaviour under similar conditions. Of course in the isotropic case $e = 1$ and $c_2 = c_3$.

For values $\Omega \ll 1$ corresponding to very large wavelengths (exceeding considerably the perimeter of the tube) the tubes behaves obviously as a rod in which axial waves under uniaxial stress conditions propagate under nondispersive conditions ($c \approx c_2$, independent of frequency). The main displacement component is u_2, radial displacements u_1 can be neglected. For increasing values of Ω, as long as $\Omega < 1$, the influence of the radial displacements becomes stronger and causes more energy to be canalized to radial motion rather than axial wave propagation. Thus the value of the phase velocity drops. For values of Ω near 1 the radial motion becomes so strong that energy transport in the axial direction becomes practically impossible in the membrane mode ("radial resonance"). Besides, since for such values the phase velocity given by (1.9) approaches zero, the corresponding wavelengths will be too small to fulfill the condition $2\pi R = O(\Lambda)$ assumed in the derivation of (1.9). On the other side, for increasing values $\Omega > 1$, the radial motion decreases and again axial displacements become more important. Since in this case the wavelengths are smaller than the perimeter (but still large with respect to the wall thickness H), the curvature of the tube ceases to be effective and the waves tend to propagate in a longitudinal nondispersive "plate mode" with negligible hoop strains and $c \approx c_2/(1 - ev^2)$. Of course this discussion follows very closely arguments well-known from isotropic theory (see for example [9]).

II) Wavelengths comparable with the geometric average of radius and thickness:

$$\Lambda = O\left(\sqrt{RH}\right) \quad \text{or} \quad \varepsilon_R = O[(\varepsilon_A)^2].$$

The asymptotic analysis in this case and under assumption (1.2) shows the following important feature which differs essentially from isotropic behaviour: In the present case, coupled with flexural stiffness in the axial direction, transverse shear deformation becomes so important that simplifying a priori assumptions on the distribution of strains and stresses across the thickness (for example linear or cubic) can no longer be justified even in a first step of approximation. The proper distribution must be determined from partial differential equations containing derivatives both with respect to x_2 and to the thickness coordinate x_1 by considering the boundary conditions on the stress-free surfaces of the tube wall. Following ideas similar to the ones developed in [6] for beams, we find for *sinusoidal waves* (details will be given in [8]) in a first step of approximation

$$u_1 = U_1(x_2, t), \qquad u_2 = U_2(x_2, t) - U_{1,2} \frac{\text{Sh}(\alpha x_1)}{\alpha \, \text{Ch}(\alpha H/2)}, \qquad (1.10)$$

where

$$\alpha := \frac{1}{\sqrt{\varepsilon_M(1-ev^2)}} \frac{2\pi}{\Lambda}. \tag{1.11}$$

The "midsurface" displacement components U_1 and U_2 in (1.10) fulfill the following differential equations

$$U_{1,22} + \left(\frac{2\pi}{\Lambda}\right)^2 U_1 = 0,$$

$$RU_{2,2} + evU_1 = 0, \tag{1.12}$$

$$R^2 e^{-1}\varepsilon_M \left[1 - \frac{\mathrm{Th}\,(\alpha H/2)}{\alpha H/2}\right] U_{1,22} - U_1 = R^2(c_3)^{-2}\,U_{1,tt}$$

which lead directly to the dispersion relation formulated in terms of the dimensionless circular frequency (1.6)

$$\Omega^2 = 1 + \varepsilon_M\,e^{-1}\left[1 - \frac{\mathrm{Th}\,(\alpha H/2)}{\alpha H/2}\right]\left(\frac{2\pi R}{\Lambda}\right)^2. \tag{1.13}$$

It is clear from this relation that the assumption about the order of magnitude of the wavelength, that is $\Lambda = O\bigl(\sqrt{RH}\bigr)$, cannot be fulfilled near $\Omega = 1$ since for this value $\Lambda^{-1} = 0$. Thus the solution above can only be used for values of Ω "sufficiently" larger than 1 (but not too large, otherwise the assumption $\Lambda \gg H$ cannot be satisfied). The function {frequency $\Omega \to$ phase velocity c} is defined in parametric form with Λ as a parameter with (1.11), (1.13) and the relation

$$c = \Lambda f = \frac{\Lambda}{2\pi R}\,\Omega c_3. \tag{1.14}$$

Obviously, for the orders of magnitude of Λ for which (1.13) remains valid, the phase velocity is quite smaller than c_3. This is not surprising since one is dealing here with flexural waves.

III) Transition regime near frequencies $f = c_3/(2\pi R)$

Since the membrane solution leading to (1.9) fails for values of Ω approaching 1 from the left side ($\Omega < 1$) and the flexural solution leading to (1.13) fails for Ω approaching 1 from the right side ($\Omega > 1$) a transition regime for values of Ω near 1 must be found to match both solutions. This is a typical problem of singular perturbation theory. It has been solved for the present case with the following result (see [8] for details) for the radial displacement $u_1(x_2, t)$ performing sinusoidal motion with frequencies Ω near 1:

$$\frac{R^4H^2}{12e(1-ev^2)}\,u_{1,222222} - (\Omega^2 - 1)\,R^2u_{1,22} + e^2v^2u_1 = 0. \tag{1.15}$$

The corresponding dispersion relation in terms of frequency and wavelength can easily be derived from (1.15) as

$$\Omega^2 = 1 - e^2v^2\left(\frac{\Lambda}{2\pi R}\right)^2 + \frac{1}{12e(1-ev^2)}\frac{(2\pi RH)^2}{\Lambda^4} \tag{1.16}$$

and must be used for values of Ω "near" 1 (both on the "left" and "right" side of 1) instead of (1.9) and (1.13).

The phase velocity follows again from (1.14), so that the function {frequency $\Omega \rightarrow$ phase velocity c} is given in parametric form (with Λ as a parameter) by (1.14) and (1.16).

2.3 Comparison of theoretical predictions with measurements

The theoretical results outlined above will be used in the following to discuss measured frequencies and wave-numbers for one out of the ten sets of results available from all tubes which were tested. Since the other nine cases produced roughly the same pattern (they will be discussed in detail in [8]) results for one of the tubes with 22.5° winding angle will be chosen as a typical example and discussed shortly below. The corresponding tube will be labelled for further reference as "tube 221".

As stated above, in the frequency range $\Omega < 1$ leading to wavelengths larger than the perimeter of the tube ($\Lambda > 2\pi R$) and labelled as "case Ia" for further reference, measuring the resonance frequencies of the tube in the first $10-70$ modes (27 modes in the case of tube 221) proved to be more efficient than transient wave evaluations. This range is covered by the dispersion relation (1.9). The latter gives also for $\Omega > 1$ the phase velocities of the fast wave in membrane mode (plate wave). The corresponding range with wavelengths $\lambda = O(2\pi R)$ will be called "case Ib" for further reference. In this case Dual measured narrow-band transient waves and obtained by proper Fourier-analysis for tube 221 a total of 108 wave-numbers corresponding to 108 frequencies. The 27 resonance frequencies of case Ia can be computed from the simplified theory leading to (1.3) and compared with the measured values. For this computation boundary conditions must be used taking into account the inertia of the transducers attached to a rigid ring fixed at one end of the tube (the other end was free). For tube 221 the total mass of ring and transducers was 0.184 times the mass of the tube. From the 108 measured wave-numbers $2\pi/\Lambda$ of case Ib and the simple dispersion relations (1.9), (1.14) the corresponding frequencies can also be computed and compared with the measured values. Knowledge of the moduli E_2, E_3 and v is necessary to perform both comparisons. These values were obtained by the following "least square fit" procedure: For a set of values $\{E_2, E_3, v\}$ the least square relative error was computed for the set of 27 frequencies of case Ia and 108 frequencies of case Ib separately. The simple (non-weighted) average of the two error values was then minimized by proper choice of $\{E_2, E_3, v\}$. The following results were obtained for tube 221:

$$E_2 = 7.52 * 10^{10} \text{ N/m}^2, \quad E_3 = 1.155 * 10^{10} \text{ N/m}^2, \quad v = 1.275 \tag{1.17}$$

(where $v := v_{23}$ is the ratio of the negative lateral strain $-\varepsilon_{33}$ obtained in the hoop direction to the axial strain ε_{22} under uniaxial stress σ_{22} parallel to the axis of the tube). With these values the set of 27 measured resonance frequencies of case Ia could be fitted with a relative variance of 0.78% and the set of 108 frequencies measured in case Ib with a variance of 0.86%.

In the short wave range $L = O(\sqrt{RH})$ leading to flexural modes of deformation Dual obtained by Fourier-analysis of the measurements 153 pairs of values {frequency, wave-number}. From the measured wave-numbers and with the help of (1.17) the frequencies in the flexural mode (1.13) labelled in the following as "case II" and in the transitional mode (1.16) labelled as "case III" can be computed, provided that in the former case an adequate value of the transverse shear modulus G_{12} in the axial direction is chosen. To obtain this value the calculated and measured frequencies were fitted to minimize the relative square error. With the value

$$G_{12} = 0.43 * 10^{10} \text{ N/m}^2 \tag{1.18}$$

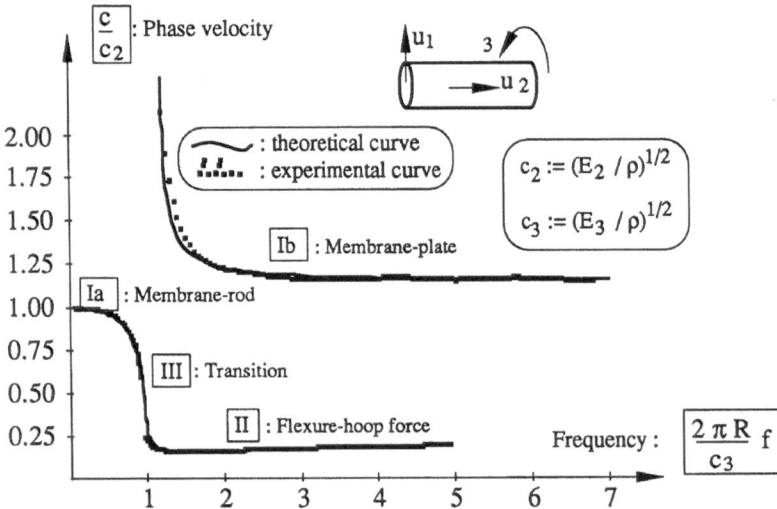

Fig. 6. Phase velocity as a function of frequency for axisymmetric structural waves in a filament-wound tube with 22.5° winding angle. The moduli E_2, E_3 and ν were found by fitting measured data in the ranges I a and I b. The transverse shear modulus G_{12} is then determined by fitting data in range II. Except at "singular" regions where the phase behaviour is difficult to measure (left side of I b and II) the match between measured and calculated functions is excellent

a minimum relative variance of 0.92% could be reached. This is a remarkable result, especially since the largest errors ranging from 6 to 3% were only found in the first 3 of the 153 frequencies in a range where the phase velocities are the lowest and thus the measurements quite difficult. For higher values of the frequency the relative error dropped to lower than 0.2%.

The theoretical values of the phase velocity calculated with the help of the moduli (1.17), (1.18) from the dispersion relations presented in the former subsection have been compared with values obtained from the experimental data in Fig. 6. The excellent fit confirms the quality of both the experiments and the theory.

3 Evaluation of global laminate stiffnesses in a composite plate by structural wave propagation

3.1 Experimental preliminaries

Consider a laminate assembled out of several unidirectionally fiber-reinforced laminae, symmetrically arranged to form a plate which behaves macroscopically orthotropic. The knowledge of its flexural stiffnesses, of its in-plane tensile stiffnesses and of its transverse and in-plane shear moduli may prove useful in many applications. For example these values may help detect possible faults in the assemblage pattern of the laminae, they may be used as input data for dynamic or static predictions on loading capacity etc. Classical procedures of measurement including strain-gauge methods may prove to be inaccurate or inadequate to provide the desired information. The technique based on structural wave propagation described below has given very satisfactory results. The experimental set-up is similar to the one used on tubes (see Figs. 2 and 3). A piezoelectric transducer capable of producing transverse (flexural) or in-plane displacements is mounted on the surface of the plate, sufficiently far from the boundaries (Fig. 4). In the case of an isotropic structure the shear transducer arrangement illustrated in Fig. 4 (four annular sectors with alternating polarities) would only produce in-plane torsional displacements

u_θ propagating with polar symmetry (independent of the polar angle θ) in the radial direction with the shear wave velocity $\sqrt{G/\varrho}$. In an orthotropic structure shear and longitudinal modes are only uncoupled in the principal direction of anisotropy. The local shear motion with the azimutal displacement component u_θ produced in the orthotropic structure by the transducer illustrated Fig. 4 is immediately transformed by mode conversion in mixed torsional and longitudinal motion. Besides, whereas in isotropic structures the curves of constant phase emanating from a point-like source are circular (i.e. of second order in the in-plane coordinates x_1, x_2), in the orthotropic structure phase curves of higher order are produced both in flexural and membrane (in-plane) modes.

The frequency band of the pulse generated with the transducer must be chosen carefully. The lower limit should be sufficiently high to avoid reflections from the plate boundaries reaching the points at which the incident wave is being measured. At the higher limit of the frequency band the corresponding wavelengths should be sufficiently large (for example at least 5 times larger than the plate thickness) to ensure that the laminate behaves as a homogeneous orthotropic plate in flexural or membrane modes. On the other hand the frequency band should not be too narrow, since one is interested in measuring frequency-dependent phase velocities for a frequency interval as wide as possible in a single experiment. The response of the plate is measured at two precisely defined radial distances r_1 and r_2 from the transducer with the help of a laser-Doppler-interferometer, applying retro-reflective bands on the plate surface at appropriate places (as in the experiments on tubes described in Sect. 1). The displacements should be measured on the side of the plate opposite to the one with the transducer, to avoid interference with possible surface waves. The smaller distance r_1 should be large enough to allow the individual frequency components of the pulse to travel according to a fairly sinusoidal pattern as a function of distance (in centrally induced waves the distance function of harmonic waves is strictly non-sinusoidal in the immediate neighbourhood of the load; it approaches a sinusoidal pattern quite rapidly with travelling distance, in most cases after one or two "wavelengths"). The larger distance r_2 should be chosen small enough to avoid interference of the incident pulse with reflected waves coming from the plate boundaries. Too small intervals $r_2 - r_1$ reduce of course the accuracy of the measured wave velocities.

As in experiments for cases Ib, II and III on the tubes, the recorded measurements are fed as digital data into the computer (see Fig. 2) and Fourier-analyzed to obtain the phase spectra at both r_1 and r_2. The phase velocity c for a given frequency f is evaluated from the phase difference $\Delta\varphi := \varphi(f, r_2) - \varphi(f, r_1)$, using the simple relation

$$c = 2\pi f\, \frac{r_2 - r_1}{\Delta\varphi}. \tag{2.1}$$

The case of centrally induced *flexural waves* in a cross-ply laminate has been analyzed in some detail in [10] where both theoretical considerations based on asymptotic analysis and experimental aspects are presented. In this connection two cross-ply plates have been tested in our laboratory by M. Veidt as part of the work leading to his doctoral thesis [11]. The plate illustrated in Fig. 4 was produced out of 9 carbon-fiber-reinforced laminae (37% epoxy volume) arranged as $(0, 90, 0, 0, 90, 0, 0, 90, 0)$ — plies with respect to the x_1-axis of the laminate. Based on the assumption of homogeneous structural behaviour which can be justified if the wavelengths are sufficiently large with respect to the thickness of the laminae (0.25 mm in the laminate mentioned above) "equivalent" tensile elasticity moduli E_α and transverse shear moduli $G_{3\alpha}$ along the *principal directions of orthotropy* x_α ($\alpha = 1$ or 2) can be defined from the corresponding global flexural and transverse shear stiffnesses. The theory developed in [10] shows that the phase velocity along either one of the principal directions is connected to the tensile modulus E_α and the transverse shear modulus $G_{3\alpha}$ through

$$c = \frac{\dfrac{\pi H}{\Lambda}\sqrt{\dfrac{E_\alpha}{3\varrho}}}{\sqrt{1 + \left(\dfrac{2\pi H}{\Lambda}\right)^2 \dfrac{E_\alpha}{10 G_{3\alpha}}}} \qquad (2.2)$$

where H is the total thickness of the plate, Λ the wavelength and ϱ the density. Thus accurate values of E_α and $G_{3\alpha}$ can be obtained by curve-fitting with experimental data generated along the principal directions of orthotropy $\alpha = 1$ or 2. The results of the curve-fitting are illustrated in Fig. 11 for the x_2-direction. The values of the moduli obtained for the laminate mentioned above were

$$E_1 = 9.22(\pm 0.28) \times 10^{10}\ \mathrm{N/m^2}, \qquad E_2 = 4.24(\pm 0.13) \times 10^{10}\ \mathrm{N/m^2},$$

$$G_{31} = 0.420(\pm 0.017) \times 10^{10}\ \mathrm{N/m^2}, \qquad G_{32} = 0.295(\pm 0.015) \times 10^{10}\ \mathrm{N/m^2}. \qquad (2.3)$$

The accuracy of the fit was thus 3% for the tensile modulus and 5% for the shear modulus.

Theoretical considerations and experimental results on *centrally induced membrane (in-plane) waves* in orthotropic cross-ply laminates are treated in some detail in [12]: A short account of the main results will be given below:

3.2 Centrally induced in-plane structural waves in cross-ply laminates

For the case illustrated in Fig. 4 and under the assumption of homogeneous orthotropic behaviour with x_α as coordinates along the principal directions of orthotropy $\alpha = 1, 2$ the basic equations of linear elastodynamics lead to the differential equations

$$C_{11} u_{1,11} + (C_{12} + G_{12}) u_{2,12} + G_{12} u_{1,22} = \varrho u_{1,tt},$$

$$G_{12} u_{2,11} + (C_{12} + G_{12}) u_{1,21} + C_{22} u_{2,22} = \varrho u_{2,tt} \qquad (2.4)$$

for the displacement components u_1, u_2 where G_{12} is the in-plane shear modulus and the stiffnesses $C_{\alpha\beta}$ are given in terms of the elasticity moduli E_α and v_{12} as

$$C_{\alpha\alpha} = m E_\alpha \quad \text{(no summation over } \alpha),$$

$$C_{12} = m v_{12} E_2, \qquad m := \left[1 - (v_{12})^2 \frac{E_2}{E_1}\right]^{-1}. \qquad (2.5)$$

The following "small" dimensionless parameters can be identified in this case:

    (1) The ratio $G_{12}/(E_1$ or $E_2)$ whose actual value can be expected to be 0.1 or less,

    (2) the coefficient of lateral contraction v_{12} whose expected value is about 0.05,

    (3) the ratio R/r_1 of the transducer radius R to the shortest distance r_1 from its center at which the incoming pulse was measured; this value was about 0.13 in the present case.

Based on these evaluations the asymptotic analysis of (2.4) leads to the following conclusions in a first step of approximation:

    (a) Two centrally induced in-plane waves propagate in the laminate each of which carries essentially only one of the displacement components u_1 or u_2 parallel to the principal directions

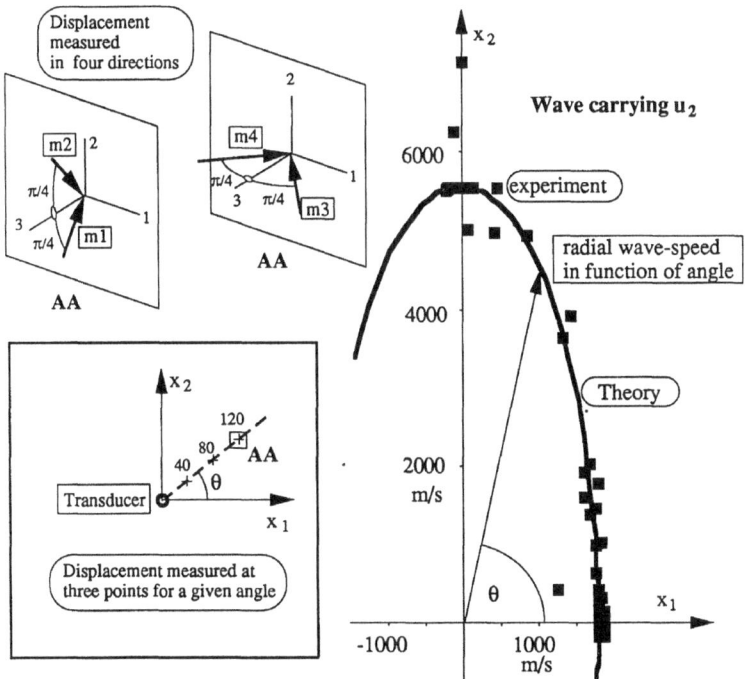

Fig. 7. Radial wave velocity as function of the angle for centrally induced waves carrying the displacement component u_2 in a cross-ply laminate. The waves were generated by piezoelectric shear transducers arranged as in Fig. 4 to produce azimutal displacements u_θ at their periphery. The cross-ply laminate with the low shear modulus reacts by producing two waves each of which carries only one displacement component parallel to one of its principal directions of orthotropy. Along direction 1 the wave carrying u_2 travels with shear wave velocity, along 2 with the corresponding longitudinal wave velocity

of orthotropy. This is very different from isotropic behaviour where mode separation in centrally induced waves occurs in torsion (azimutal displacement u_θ) and longitudinal motion (radial component u_r). Whereas in an isotropic plate a torsional transducer such as the one sketched in Fig. 4 would produce only a shear wave propagating radially with circular phase curves and carrying only one displacement component u_θ, in orthotropic laminates both waves mentioned above will be generated.

(b) The radial wave velocities $c_r(u_1)$ and $c_r(u_2)$ of both waves in the laminate depend on the angle θ between the radius along which $c_r(u_\alpha)$ is defined and the principal direction 1 (Fig. 7). This is also different from isotropic behaviour where of course both shear and longitudinal waves travel with circular symmetry. Provided that the radial distance r from the transducer center is sufficiently large with respect to the transducer radius R, the θ-dependence of the radial velocities for the two waves in the laminate is as follows:

$$c_r(u_1) = \sqrt{\frac{G_{12}E_1}{\varrho(G_{12}\cos^2\theta + E_1\sin^2\theta)}}, \tag{2.6}$$

$$c_r(u_2) = \sqrt{\frac{G_{12}E_2}{\varrho(E_2\cos^2\theta + G_{12}\sin^2\theta)}}. \tag{2.7}$$

Clearly the wave carrying the displacement u_1 propagates along the axis 1 ($\theta = 0$) with the wave velocity of axial waves $\sqrt{E_1/\varrho}$ and along the axis 2 ($\theta = \pi/2$) with the wave velocity of shear waves $\sqrt{G_{12}/\varrho}$. The roles are reversed for the wave carrying the displacement u_2.

During his work leading to his "Diploma Thesis" at the Department of Mechanical Engineering of our University, A. Eisenhut measured the radial velocities of centrally induced in-plane waves with the set-up illustrated in Figs. 2, 3 on the plate sketched in Fig. 4. For various angles θ ranging from $-5°$ to $+91°$ he measured the arrival time values and the amplitudes of narrow-band signals (Fig. 1) at three radial distances $r = 40, 80$ and 120 mm from the center of the torsional transducer which generated the signals. In order to obtain the displacement components u_1 and u_2, at each of the three distances just mentioned, he measured the displacement in 4 different directions m_1 to m_4 (see Fig. 7). Directions m_1, m_2 were chosen (corresponding prisms and retro-refractive bands were used in conjunction with the interferometer sketched in Fig. 3) in the plane 23 perpendicular to the plate, at angles $\pm 45°$ with respect to the perpendicular 3. Directions m_3 and m_4 were in the perpendicular plane 13 and built also angles of $\pm 45°$ with respect to the perpendicular 3. The desired displacement components were then obtained by simple operations (addition, subtraction, division with $\sqrt{2}$) from the measured ones.

Figure 7 shows the radial velocities (group velocity $\approx$ phase velocity) for the wave carrying the displacement u_2 and Fig. 8 for the wave carrying u_1. The experimental values of c_r were obtained from measured arrival times at the three distances mentioned above and in function of various angles θ. Thus for each θ two velocity values were available. The measured values of $c_r(u_2)$ and $c_r(u_1)$ were then compared with the theoretical results (2.7), (2.6) by proper choice of E_1, E_2 and G_{12} minimizing the relative variance of the differences between measured and calculated velocities for all angles. A relative variance of 12% was obtained for $c_r(u_2)$ obtained from 40 and 80 mm distances. The relative variance is 4.5% for the values obtained from 80 and 120 mm. For $c_r(u_1)$ the corresponding relative variances were 3.3% (for the smaller distances) and 8.6%. Of course the fit was much better for the angles at which the waves were slower (near the axis 1 for the wave carrying u_2 and near the axis for the wave carrying u_1), since the corresponding time differences were of the order of 20 µs for the slower wave and about $5 - 7$ µs for the faster one. The values of the "in-plane" moduli of elasticity obtained by the procedure above are:

$$E_1 = 9.3 \times 10^{10} \ \text{N/m}^2, \qquad E_2 = 4.7 \times 10^{10} \ \text{N/m}^2, \qquad G_{12} = 0.48 \times 10^{10} \ \text{N/m}^2. \tag{2.8}$$

The fit with values (2.3) derived from experiments with flexural waves is excellent for the first modulus and fair for the second one (of course the in-plane shear modulus G_{12} is expected to be

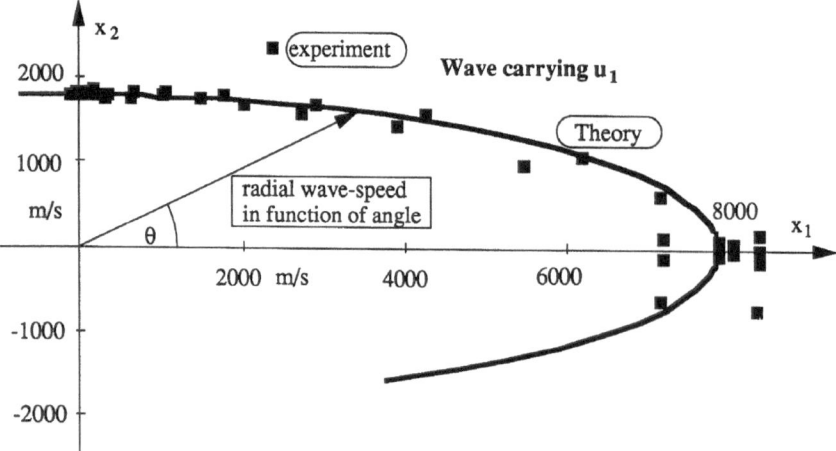

Fig. 8. Radial wave velocity as function of the angle for centrally induced waves carrying the displacement component u_1 in a cross-ply laminate

different from the transverse shear moduli). Even in the case of E_2 the two results which differ by 10% can be considered to be consistent since in orthotropic laminates the laminate theory shows that in-plane moduli are not necessarily equal to moduli derived from global flexural stiffnesses.

References

[1] K. Graff: Wave motion in elastic solids. Ohio University Press 1975.

[2] J. D. Achenbach: Wave propagation in elastic solids. North-Holland 1973.

[3] J. Miklowitz: Elastic waves and waveguides. North-Holland 1978.

[4] Review of progress in quantitative nondestructive evaluation (D. O. Thompson, D. E. Chimenti, eds.), Vols. 8 A, 8 B. Plenum Press 1989.

[5] Elastic waves and ultrasonic nondestructive evaluation, S. K. Datta, J. D. Achenbach, Y. S. Rajapakse (eds.) (Proceedings of the IUTAM Symposium of July 30 – August 3, 1989). North Holland 1990.

[6] M. B. Sayir: Flexural vibrations of strongly anisotropic beams. Ingenieur Archiv 49, 309 – 323 (1980).

[7] J. Dual: Experimental methods in wave propagation in solids and dynamic viscometry. Diss. ETH 8659, Zurich 1989.

[8] M. Sayir, J. Dual, B. Gasser: Structural axisymmetric waves in anisotropic cylindrical shells: Asymptotic theory and experiments (to be published).

[9] J. H. Heimann, H. Kolsky: The propagation of elastic waves in thin cylindrical shells. J. Mech. Phys. Solids 14, 121 – 130 (1966).

[10] M. Veidt, M. Sayir: Experimental evaluation of global composite laminate stiffnesses by structural wave propagation. J. Composite Materials 24, 688 – 706 (1990).

[11] M. Veidt: Studien zum bruchmechanischen und strukturdynamischen Verhalten von Faserverbundplatten. Diss. ETH 9424, Zürich 1991.

[12] M. Sayir, M. Veidt, A. Eisenhut: Centrally induced in-plane structural waves in cross-ply laminates: Asymptotic theory and experiments (to be published).

Author's address: Prof. M. B. Sayir, Institute of Applied Mechanics, Swiss Federal Institute of Technology (ETH Zentrum), CH - 8092 Zurich, Switzerland.

Acta Mechanica (1992) [Suppl] 3: 23–34

Forced vibration of strongly-coupled structural systems connected by sub-structures

K.-W. Min, Seoul, Korea, and **T. Igusa** and **J. D. Achenbach**, Evanston, Illinois, U.S.A.

Summary. A novel method is presented to analyze the harmonic forced vibration of several main structures connected by sub-structures. Lagrange's equations are used to develop equations of motion in terms of impedances, mobilities, and modal forces. Then, a frequency window method is used to reduce the complexity of the problem. The window contains resonance terms which are analyzed in detail. The remaining terms outside the window are slowly varying with respect to the forcing frequency and are analyzed with less detail. This approach is efficient because computational effort is concentrated on the most important terms in the response. With the same computational effort, it is possible to include more total modes in the frequency window method than in the standard mode truncation approach.

1 Introduction

Complicated structural systems are found in mechanical, civil and aerospace engineering. Although finite elements and mode synthesis can be used to compute the dynamic response of such systems at low to mid frequencies, the structural complexity has made the interpretation of the results of such analyses difficult. Many researchers in dynamic analysis have sought to further develop existing theories or develop alternate methods to obtain greater insight in the behavior of connected sub-structures. Gladwell and Bishop [1], Nicholson and Bergman [2], and Kelkel [3] used Green's functions of vibrating continuous structures; Jacquot and Soedel [4] used modal expansions and sub-structure impedances to formulate the equations of motion; Hallquist and Snyder [5] used "tie-vectors" to develop equations for combined sub-structures; Leissa [6] investigated curve-veering phenomena in coupled modes; Dowell [7] used Lagrange multipliers for the constraints at the interface of connected structures; and Sackman and Kelly [8], Sackman et al. [9] and Igusa and Der Kiureghian [10] have combined mode synthesis concepts, matrix algebraic theory, and perturbation methods for characterizing weakly-coupled sub-structures.

Recently, Igusa, Achenbach, and Min [11, 12] have developed a new technique to analyze the dynamic characteristics of coupled, continuous substructures. Their analysis begins with the eigenvalue problem for the combined system which is derived from Lagrange's equations and is recast in terms of modal expansions of the sub-structure impedances and the main structure mobility. The central concept of the technique is a frequency window method wherein frequency dependence is maintained for frequency terms which lie within a range of interest. Terms outside the window are also retained, but without the frequency dependence.

The frequency window method is fundamentally different from mode truncation. Mode truncation includes terms associated with all modes whose natural frequencies are less than some cut-off frequency. Since frequency dependence is included in all terms, the computational complexity is related to the total number of retained modes. The frequency window method also needs mode truncation to make the number of terms finite. However, the terms associated with modes whose natural frequencies lie outside a window about the excitation frequency are nearly constant with respect to small variations of the excitation frequency. These terms are not truncated, but are retained as constant values evaluated at a fixed value of the excitation frequency within the frequency window. Thus, the computational complexity is related to the number of modes whose natural frequencies lie within the frequency window. This means that, with the same computational effort, it is possible to include more total modes in the frequency window method than in the standard mode truncation approach.

In this paper, the frequency window method is used to analyze harmonic forced vibration of several main structures connected by sub-structures. The window contains resonance terms which are analyzed in detail. The remaining terms outside the window are slowly varying with respect to the forcing frequency and can be analyzed with less detail. This approach is efficient because computational effort is concentrated on the most important terms in the response.

2 Formulation

Consider the system illustrated in Fig. 1 consisting of $n + 1$ main structures with domains $x \in \Omega_{(l)}$, $l = 1, 2, \ldots, n + 1$ and n sub-structures with domains $x \in \Omega_k$, $k = 1, 2, \ldots, n$. Each sub-structure k is rigidly connected to the two neighboring main structures at the sub-structure boundary points, $x_k = \{x_{k1}, x_{k2}\}$. (Throughout the paper, vectors and matrices are denoted by bold lower case letters and bold upper case letters, respectively.) An external force, $qe^{-i\omega t}$, is applied at points x_f of the l-th main structure, where $i = \sqrt{-1}$ (which is not italic, to distinguish it from the subscript i).

The dynamic properties of the l-th main structure with free boundaries are described by mode shapes, $\phi_{(l)i}(x)$, and natural frequencies, $\omega_{(l)i}$, where $x \in \Omega_{(l)}$ is the coordinate vector, i is the mode number, and the mode shapes are normalized with respect to the mass density. Similarly, the dynamic properties of each sub-structure k with fixed boundaries are described by normalized mode shapes, $\psi_{kj}(x)$, natural frequencies, ω_{kj}, and mass density, $\varrho_k(x)$, where $x \in \Omega_k$ and j is the mode number. (For clarity, the subscripts i and j will generally be associated with main structures and sub-structures, respectively, throughout the paper.)

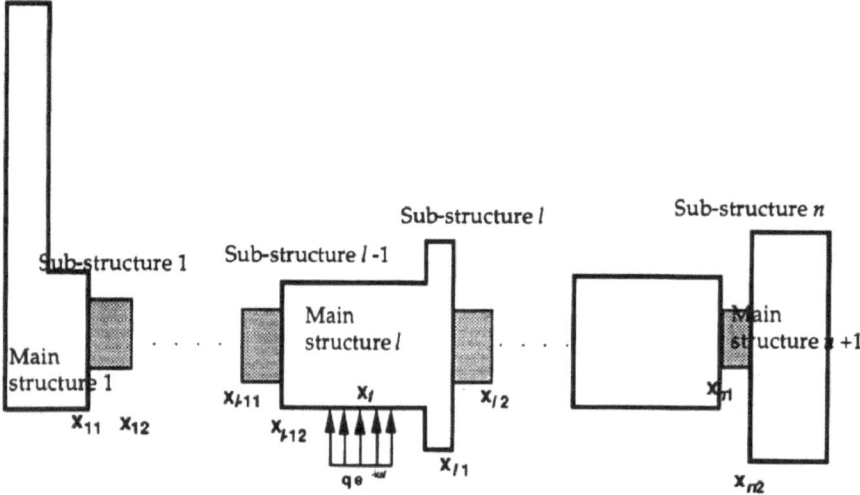

Fig. 1. Main structures connected by a sub-structures with external force

It is also necessary to introduce functions describing the static response characteristics of each sub-structure k. Each sub-structure has four boundary coordinates: translation and rotation at the left boundary x_{k1}, and translation and rotation at the right boundary, x_{k2}. Let the 4-vector $p_k(x)$ be the static displacement function in which the i-th element corresponds to the continuous displacement of sub-structure k resulting from a unit displacement at boundary at coordinate i with the other boundary coordinates fixed. Let the 4×4 matrix S_k be the static stiffness matrix for sub-structure k in which the (i, j)-th element of S_k corresponds to the reaction force at boundary coordinate i resulting from unit force applied at boundary coordinate j with other boundary coordinates fixed.

In the following, Lagrange's equations are used to develop equations of motion for the combined structure in terms of $\phi_{(l)i}(x)$, $\psi_{kj}(x)$, $\omega_{(l)i}$, ω_{kj}, $p_k(x)$, and S_k. This equation is subsequently analyzed by the frequency window method to determine the response characteristics of the combined structure.

The forced vibration displacement field for the l-th main structure is given by

$$w_l(x, t) = \sum_{i=1}^{\infty} a_{(l)i}\phi_{(l)i}(x)\, e^{-i\omega t} \tag{1}$$

where $x \in \Omega_{(l)}$ and $a_{(l)i}$ are modal amplitudes for the displacements. The displacement field for the k-th sub-structure is

$$u_k(x, t) = \left[\sum_j b_{kj}\psi_{kj}(x) + f_k^T p_k(x) \right] e^{-i\omega t} \tag{2}$$

where $x \in \Omega_k$, b_{kj} are modal amplitudes for the displacements, and f_k is the 4-vector of the boundary displacements. In the following, the omnipresent harmonic term $e^{-i\omega t}$ will not be explicitly included in the equations.

The Lagrange equation formulation (Dowell [7]) is used to combine the equations of motion. The kinetic energy, K, and potential energy, U, are given by

$$K = \frac{1}{2}\,\omega^2 \left\{ \sum_{l=1}^{n+1} \sum_{i=1}^{\infty} a_{(l)i}^2 + \sum_{k=1}^{n} \left[\sum_{j=1}^{\infty} b_{kj}^2 + 2\sum_{j=1}^{\infty} b_{kj}f_k^T m_{kj} + f_k^T M_{kpp} f_k \right] \right\}, \tag{3}$$

$$U = \frac{1}{2} \left\{ \sum_{l=1}^{n+1} \sum_{i=1}^{\infty} a_{(l)i}^2 \omega_{(l)i}^2 + \sum_{k=1}^{n} \sum_{k=1}^{\infty} b_{kj}^2 \omega_{kj}^2 + \sum_{k=1}^{n} f_k^T S_k f_k \right\}. \tag{4}$$

For each sub-structure k,

$$M_{kpp} \equiv \int_{\Omega_k} p_k(x)\, p_k^T(x)\, \varrho_k(x)\, dx, \tag{5}$$

is a 4×4 modal mass matrix in which the (i, j)-th element of M_{kpp} represents the coupling between the i-th and j-th elements of $p_k(x)$. (The product in the integrand is an outer product between a column vector and a row vector.) Also, for each substructure k,

$$m_{kj} \equiv \int_{\Omega_k} p_k(x)\, \psi_{kj}(x)\, \varrho_k(x)\, dx, \tag{6}$$

is a 4-component modal mass vector in which the i-th element of m_{kj} represents the coupling between the i-th element of $p_k(x)$ and the j-th mode shape of sub-structure k. The physical dimensions of M_{kpp} and m_{kj} are mass and $(\text{mass})^{1/2}$, respectively, due to the mass normalization of the mode shape $\psi_{kj}(x)$.

The constraints of the system are specified by the displacements and rotations at the boundary points x_k. Equating the displacements of sub-structure k with those of the main structures yields

$$f_k = \sum_{i=1}^{\infty} \begin{bmatrix} a_{(k)i}\, \phi_{(k)i}(x_{k2}) \\[6pt] a_{(k)i}\, \dfrac{\partial \phi_{(k)i}}{\partial x}(x_{k2}) \\[6pt] a_{(k+1)i}\, \phi_{(k+1)i}(x_{k1}) \\[6pt] a_{(k+1)i}\, \dfrac{\partial \phi_{(k+1)i}}{\partial x}(x_{k1}) \end{bmatrix}. \tag{7}$$

Using a 4-vector of Lagrange multipliers, λ_k, for the displacement constraints of each sub-structure k in Eq. (7), the Lagrange equations of motion are

$$(\omega_{(l)i}^2 - \omega^2)\, a_{(l)i} + \lambda_{l-1,3}\phi_{(l)i}(x_{l1}) + \lambda_{l-1,4}\frac{\partial \phi_{(l)i}}{\partial x}(x_{l1}) + \lambda_{l,1}\phi_{(l)1}(x_{l2}) + \lambda_{l,2}\frac{\partial \phi_{(l)i}}{\partial x}(x_{l2}) = 0 \tag{8}$$

for $l = 1, \ldots, n+1$, and

$$(\omega_{kj}^2 - \omega^2)\, b_{kj} - \omega^2 f_k^T M_{kj} = 0 \tag{9}$$

$$\omega^2 \left\{ M_{kpp} f_k + \sum_j b_{kj} M_{kj} \right\} - S_k^T f_k + \lambda_k = 0 \tag{10}$$

for $k = 1, \ldots, n$. Here, λ_{kj} are the components of the 4-vector λ_l, and λ_0 and λ_{n+1} are defined to be vectors of zeros.

The Lagrange Eqs. (8)–(10) are rewritten by using the support motions f_k as the unknown variables. First, Eq. (9) is used to express b_{kj} in terms of f_k

$$b_{kj} = \frac{\omega^2}{\omega_{kj}^2 - \omega^2}\, m_{kj}^T f_k. \tag{11}$$

Substituting b_{kj} in Eq. (10) and solving for the Lagrange multipliers results in expressions in terms of f_k. Combining the result with Eq. (8) yields expressions for $a_{(l)i}$ in terms of f_k as

$$a_{(1)i} = \frac{i\omega}{\omega_{(1)i}^2 - \omega^2}\, \hat{\Phi}_{(1)i}^T(x_{11})\, Z_1 f_1 \tag{12}$$

$$a_{(l)i} = \frac{i\omega}{\omega_{(l)i}^2 - \omega^2}\, \left\{ \hat{\Phi}_{(l)i}^T(x_{l-1,2})\, Z_{l-1} f_{l-1} + \hat{\Phi}_{(l)i}^T(x_{l1})\, Z_l f_l \right\}$$

for $l = 2, 3, \ldots, n$ \hfill (13)

$$a_{(n+1)i} = \frac{i\omega}{\omega_{(n+1)i}^2 - \omega^2}\, \hat{\Phi}_{(n+1)i}^T(x_{n2})\, Z_n f_n. \tag{14}$$

Here,

$$Z_k(\omega) = -i\omega \left\{ M_{kpp} - \frac{S_k}{\omega^2} + \sum_j \frac{\omega^2}{\omega_{kj}^2 - \omega^2}\, M_{kj} M_{kj}^T \right\} \tag{15}$$

is the 4×4 impedance matrix of sub-structure k in which the (i, j)-th element of $Z_k(\omega)$ corresponds to the reation force divided by $e^{-i\omega t}$ at boundary coordinate i resulting from a velocity $e^{-i\omega t}$

applied at boundary coordinate j with all other boundary coordinates fixed. Also, $\hat{\boldsymbol{\Phi}}_{(k)i}(x)$ is the displacement and rotation of mode i of main structure k at x.

$$
\hat{\boldsymbol{\Phi}}_{(k)i}(x_{k1}) \equiv \begin{bmatrix} \phi_{(k)i}(x_{k1}) \\ \dfrac{\partial \phi_{(k)i}}{\partial x}(x_{k1}) \\ 0 \\ 0 \end{bmatrix}, \quad \hat{\boldsymbol{\Phi}}_{(k)i}(x_{k-1.2}) \equiv \begin{bmatrix} 0 \\ 0 \\ \phi_{(k)i}(x_{k-1,2}) \\ \dfrac{\partial \phi_{(k)i}}{\partial x}(x_{k-1,2}) \end{bmatrix}. \tag{16}
$$

Equations (11)–(14) give the motion of the sub-structures and the main structure, respectively, in terms of the motions f_k at the boundary points. The support motions (i.e. displacements and rotations) are obtained by substituting Eq. (7) into Eqs. (11)–(14) and switching the order of summation. The result is

$$
[\boldsymbol{I} + \boldsymbol{N}(\omega)\,\mathbf{diag}\,\{\boldsymbol{Z}_1(\omega), \dots, \boldsymbol{Z}_n(\omega)\}]\,\boldsymbol{f} = \boldsymbol{Q}(\omega) \tag{17}
$$

where $\boldsymbol{I}$ is the $4n \times 4n$ identity matix, f is the $4n$-support displacement vector

$$
\boldsymbol{f} = [\boldsymbol{f}_1^T, \dots, \boldsymbol{f}_n^T]^T \tag{18}
$$

and $\mathbf{diag}\,\{.\}$ represents a block diagonal matrix with diagonally aligned sub-matrices listed in the braces. The modal force vector $\boldsymbol{Q}(\omega)$ is defined by $\boldsymbol{Q}(\omega) \equiv [0, 0, \boldsymbol{Q}_{(2)}^T \dots \boldsymbol{Q}_{(n)}^T, 0, 0]^T$ where

$$
\boldsymbol{Q}_{(l)}(\omega) = \sum_i \frac{q_i}{\omega_{(l)i}^2 - \omega^2}\,\boldsymbol{\Phi}_{(l)i}. \tag{19}
$$

Here, q_i are modal force coefficients given in terms of the applied force $q(x)$ which, for notational simplicity, is assumed to act only on main structures 2 through n

$$
q_i = \int_{\Omega_{(l)}} \phi_{(l)i}(x)\,q(x)\,dx \tag{20}
$$

and $\boldsymbol{\Phi}_{(l)i} = \hat{\boldsymbol{\Phi}}_{(l)i}(x_{l1}) + \hat{\boldsymbol{\Phi}}_{(l)i}(x_{l-1,2})$. Finally, $\boldsymbol{N}(\omega)$ is the mobility matrix of the main structures

$$
\boldsymbol{N}(\omega) \equiv \begin{bmatrix} \boldsymbol{N}_{11}(\omega) & \boldsymbol{N}_{12}(\omega) & & & 0 \\ \boldsymbol{N}_{21}(\omega) & \boldsymbol{N}_{22}(\omega) & \boldsymbol{N}_{23}(\omega) & & \\ & \boldsymbol{N}_{32}(\omega) & \boldsymbol{N}_{33}(\omega) & & \boldsymbol{N}_{34}(\omega) \\ & & \ddots & & \ddots \\ & & \boldsymbol{N}_{n-1,n-2}(\omega) & \boldsymbol{N}_{n-1,n-1}(\omega) & \boldsymbol{N}_{n-1,n}(\omega) \\ 0 & & & \boldsymbol{N}_{n,n-1}(\omega) & \boldsymbol{N}_{nn}(\omega) \end{bmatrix}. \tag{21}
$$

The mobility matrix $\boldsymbol{N}(\omega)$ consists of the mobilities and cross mobilities of the n main structures at the $2n$ support points $x_{11}, x_{12}, \dots, x_{n2}$. The submatrix $\boldsymbol{N}_{kk}(\omega)$ is the mobility at x_{k1} and x_{k2}

$$
\boldsymbol{N}_{kk}(\omega) \equiv \sum_i \frac{-i\omega}{\omega_{(k)i}^2 - \omega^2}\,\hat{\boldsymbol{\Phi}}_{(k)i}(x_{k1})\,\hat{\boldsymbol{\Phi}}_{(k)i}^T(x_{k1}) + \sum_i \frac{-i\omega}{\omega_{(k+1)i}^2 - \omega^2}\,\hat{\boldsymbol{\Phi}}_{(k+1)i}(x_{k2})\,\hat{\boldsymbol{\Phi}}_{(k+1)i}^T(x_{k2}) \tag{22}
$$

and $\boldsymbol{N}_{k,k+1}(\omega)$ is the cross mobility at x_{k2} and $x_{k+1,1}$

$$
\boldsymbol{N}_{k,k+1}(\omega) \equiv \sum_i \frac{-i\omega}{\omega_{(k+1)i}^2 - \omega^2}\,\hat{\boldsymbol{\Phi}}_{(k+1)i}(x_{k+1,1})\,\hat{\boldsymbol{\Phi}}_{(k+1)i}^T(x_{k2}). \tag{23}
$$

In summary, the harmonic response problem obtained through Lagrange equations has been reformulated in analytical form in terms of the mobilities of the main structure and impedances of the sub-structures. The result, given by Eq. (17), is in a rational matrix polynomial form which is similar to the results by Jacquot and Soedel [4] and Dowell [7]. This form of the harmonic response equations is not commonly used in structural dynamics; however, it has certain mathematical properties which facilitate further analytical treatment. In the following sections, Eq. (17) will be used to examine system complexity and the characteristics of sub-structure resonances and coupling using a frequency window method developed by Igusa, Achenbach, and Min [11, 12].

3 Frequency window method

The results of the preceding section are exact only when infinite numbers of modes are used. Since this is computationally impossible, modal truncation is used, where terms associated with modes whose natural frequencies are less than some cut-off frequency are retained.

The complexity of the resulting problem can be measured by the polynomial degree of Eq. (17). If there are $P(l)$ modes in main structure l and $Q(k)$ modes in sub-structure k, then the complexity Θ of the equation is

$$\Theta = \sum_l P(l) + \sum_k Q(k). \tag{24}$$

The standard method of reducing complexity is to reduce the number of modes $P(l)$ and $Q(k)$ used in the analysis. Another approach, which is used herein, is to separate dominant and non-dominant terms in the response equation and to make approximations in the non-dominant expressions.

Consider the harmonic response for a range of frequencies in the neighborhood of a central frequency, ω_0. On examining the summations in Eqs. (15), (19), (22), and (23), it can be seen that certain terms become relatively large when the natural frequencies of the structures are close to ω_0 or equivalently, in a neighborhood of ω_0. Sets of indices can be defined which correspond to these structure modes

$$I(\omega_0, \delta, l) \equiv \{\text{all } i \text{ such that } |\omega_{(l)i} - \omega_0| < \delta/2\} \tag{25}$$

$$I(\omega_0, \delta, k) \equiv \{\text{all } kj \text{ such that } |\omega_{kj} - \omega_0| < \delta/2\} \tag{26}$$

where δ is the width of the frequency window. Here, indices $(l)i$ refer to main structure l and double indices kj refer to sub-structure k. Using this classification of structure modes, the rational polynomials of mobilities, impedances, and modal forces are decomposed into a sum of two rational polynomials:

$$N_{kl}(\omega) \approx N_{kl,0}(\omega, \omega_0, \delta) + N_{kl,1}(\omega_0, \delta) \tag{27}$$

$$Z_k(\omega) \approx Z_{k0}(\omega, \omega_0, \delta) + Z_{k1}(\omega_0, \delta) \tag{28}$$

$$Q_{(l)}(\omega) \approx Q_{(l)0}(\omega, \omega_0, \delta) + Q_{(l)1}(\omega_0, \delta). \tag{29}$$

The first function in each sum contains terms with poles within the frequency window, as defined in Eqs. (25) and (26). Since such terms may be sensitive to small variations in the excitation frequency, ω, no approximations are made. The second function in each term contains terms with poles outside the frequency window. Since such terms are insensitive to small variations in ω, they are evaluated once at the central frequency, ω_0. Thus, when evaluating responses for a range of frequencies near ω_0, only the first functions in Eqs. (27)$-$(29) must be re-evaluated at each frequency.

The mathematical expressions for the functions in Eqs. (27)$-$(29) are

$$N_{kk,0}(\omega,\,\omega_0,\,\delta) = \sum_{i \in I(\omega_0,\delta,k)} \frac{-i\omega_0}{\omega_{(k)i}^2 - \omega^2} \,\hat{\boldsymbol{\Phi}}_{(k)i}(\boldsymbol{x}_{k1})\,\hat{\boldsymbol{\Phi}}_{(k)i}^T(\boldsymbol{x}_{k1})$$

$$+ \sum_{i \in I(\omega_0,\delta,k+1)} \frac{-i\omega_0}{\omega_{(k+1)i}^2 - \omega^2} \,\hat{\boldsymbol{\Phi}}_{(k+1)i}(\boldsymbol{x}_{k2})\,\hat{\boldsymbol{\Phi}}_{(k+1)i}^T(\boldsymbol{x}_{k2}) \qquad (30)$$

$$N_{k,k+1,0}(\omega,\,\omega_0,\,\delta) \equiv \sum_{i \in I(\omega_0,\delta,k+1)} \frac{-i\omega_0}{\omega_{(k+1)i}^2 - \omega^2} \,\hat{\boldsymbol{\Phi}}_{(k+1)1i}(\boldsymbol{x}_{k+1,1})\,\hat{\boldsymbol{\Phi}}_{(k+1)i}^T(\boldsymbol{x}_{k2}) \qquad (31)$$

$$\boldsymbol{Z}_{k0}(\omega,\,\omega_0,\,\delta) = \sum_{j \in I(\omega_0,\delta,k)} \frac{-i\omega_0{}^3}{\omega_{kj}^2 - \omega^2} \,\boldsymbol{M}_{kj}\boldsymbol{M}_{kj}^T \qquad (32)$$

$$\boldsymbol{Q}_{(l)0}(\omega,\,\omega_0,\,\delta) = \sum_{i \in I(\omega_0,\delta,l)} \frac{q_i}{\omega_{(l)i}^2 - \omega^2} \,\boldsymbol{\Phi}_{(l)i} \qquad (33)$$

$$N_{kk,1}(\omega_0,\,\delta) = \sum_{i \notin I(\omega_0,\delta)} \frac{-i\omega_0}{\omega_{(k)i}^2 - \omega_0{}^2} \,\hat{\boldsymbol{\Phi}}_{(k)i}(\boldsymbol{x}_{k1})\,\hat{\boldsymbol{\Phi}}_{(k)i}^T(\boldsymbol{x}_{k1})$$

$$+ \sum_{i \notin I(\omega_0,\delta)} \frac{-i\omega_0}{\omega_{(k+1)i}^2 - \omega_0{}^2} \,\hat{\boldsymbol{\Phi}}_{(k+1)i}(\boldsymbol{x}_{k2})\,\hat{\boldsymbol{\Phi}}_{(k+1)i}^T(\boldsymbol{x}_{k2}) \qquad (34)$$

$$\boldsymbol{Z}_{k1}(\omega_0,\,\delta) = -i\omega_0 \left\{ \boldsymbol{M}_{kpp} - \frac{\boldsymbol{S}_k}{\omega_0{}^2} + \sum_{j \notin I(\omega_0,\delta,k)} \frac{\omega_0{}^2}{\omega_{kj}^2 - \omega_0{}^2} \,\boldsymbol{M}_{kj}\boldsymbol{M}_{kj}^T \right\} \qquad (35)$$

$$\boldsymbol{Q}_{(l)1}(\omega_0,\,\delta) = \sum_{i \notin I(\omega_0,\delta)} \frac{q_i}{\omega_{(l)i}^2 - \omega^2} \,\boldsymbol{\Phi}_{(l)i} \qquad (36)$$

Substituting Eqs. (30)$-$(36) into Eq. (17) yields rational polynomials with lower order than the original equation, and hence reduced complexity.

Although a reduction of complexity has been achieved, there is also a corresponding decrease in accuracy. The parameter δ, which is the width of the frequency window that is used in the definitions of the index sets (Eqs. (25) and (26)), is directly related to both complexity and accuracy. Smaller values of δ lead to smaller index sets which reduce the problem complexity. However, the reduced number of terms in Eqs. (30)$-$(33) results in a decreased acuracy. Conversely, larger δ leads to increased complexity and accuracy. The limiting case $\delta \to \infty$ results in the original complexity in Eq. (24).

A general method of determining δ is not developed in this paper. However, the example in the next section shows how δ and ω_0 can be chosen to estimate errors and increase the accuracy of the analysis results.

In summary, a method for reduction of complexity of the eigenvalue problem has been developed which maintains accuracy for terms with poles near the excitation frequency, and makes an approximation for all other terms. Herein, this method is called the frequency window method.

The frequency window method is fundamentally different from mode truncation. Mode truncation computes all terms in the modal expansions with equal accuracy. The frequency window method makes approximations for terms which are insensitive to small variations of the excitation frequency. The computational savings that is gained from this approximation can be used to include more modes in the modal expansions. This may be useful since studies have shown that very large numbers of modes are sometimes necessary to achieve convergence (Meirovich and Kwak [13]).

4 Example studies

The structure shown in Fig. 2 consists of two main beams and a sub-beam with each ends rigidly connected to the main beams. All beams are modeled as Bernoulli-Euler beams. The modulus of elasticity, E, and the loss factor, $\eta = 0.01$, are the same for all beams, and the moment of inertia, I, and the mass per unit length, m, are the same for all main beams. The moments of inertia and masses per unit length are denoted by I_s and m_s for the sub-beam. For simplicity, vibrations corresponding to axial deformations are not considered. The parameters l_s and l_2 are used to denote the lengths of the sub-beam and the second main beam, normalized by the length of the first main beam.

Two nondimensional parameters are introduced: a mass density ratio $\lambda \equiv m_s/m$ and a rigidity ratio $\tau \equiv I_s/(I\lambda)$. In the example studies, two mass density ratios, $\lambda = 3.0$ and 10.0, are considered while τ is held constant at 0.25. An external point force is applied at a distance of 2.0 from the left end of the structure, and for each example the responses at the joints are examined.

The harmonic responses are obtained using a 50-mode expansion and the frequency window method, is evaluated using 10 modes of the sub-beam and 20 modes of each main beam. Every term is evaluated at every frequency. It was found that additional terms do not produce any noticable change in the results. The same 50 modes are also used in the frequency window method. Terms associated with these modes are evaluated when the excitation frequency ω is equal to ω_0. For $\omega \neq \omega_0$, only terms associated with modes within a frequency window, as defined by Eqs. (25) and (26), are re-evaluated. Thus, the 50-mode expansion and the frequency window method will yield identical results when $\omega = \omega_0$, and the frequency window method will show small but increasing errors as the difference between ω and ω_0 increases.

In a given range of excitation frequencies, several values of ω_0 can be used, where each ω_0 value corresponds to a different frequency window. By making the separation between successive values of ω_0 small, the excitation frequency is never far from one of the ω_0 values. At the mid-point between two successive values of ω_0, the computation changes from one frequency window to another. This leads a jump in the response curve plotted versus frequency. This jump is a measure of the error in the frequency window calculations and can be used to determine appropriate spacing of ω_0 and widths δ of the frequency window.

For the first example, the following nondimensional lengths are chosen: $l_2 = 1.188$, and $l_s = 0.225$. The harmonic response of the structure is determined for forcing frequencies in the neighborhood of the fifth natural frequency of the first main beam, $\omega_{5,1}$, and the 6th natural

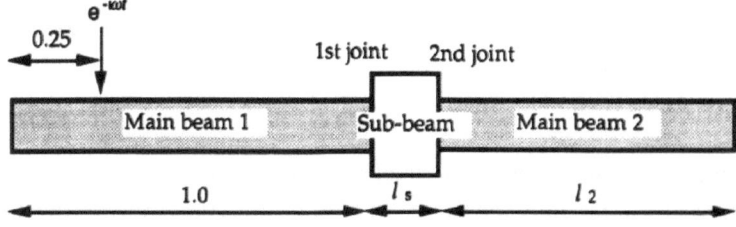

Fig. 2. Main beams connected by a sub-beam with external force

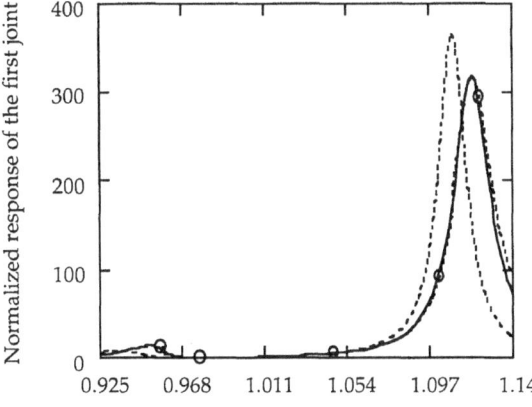

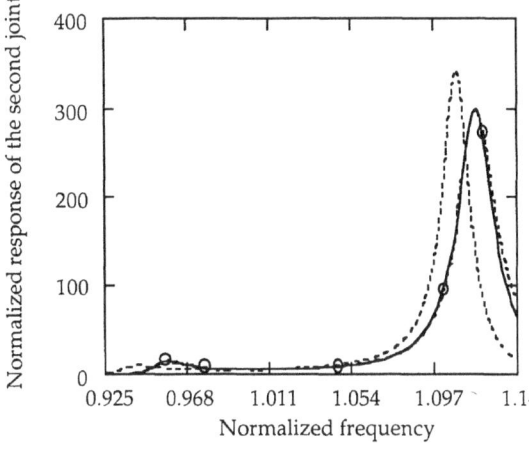

Fig. 3. Harmonic translation response for $\lambda = 3.0$ by 50-mode expansion (———) and frequency window method with $\delta/\omega_a = 0.25$ and $\omega_0 = \omega_a$, (-----) and $\delta/\omega_a = 0.52$ and nine values of ω_0 (- -○- -)

Table 1. The normalized natural frequencies of main beams and sub-beam of the first example

Mode	Main beam 1 ω_{1i}/ω_a	Sub-beam ω_j/ω_a	Main beam 2 ω_{2i}/ω_a
1	0.075	0.744	0.053
2	0.208	2.050	0.147
3	0.407	4.019	0.289
4	0.673	6.643	0.477
5	1.005	9.924	0.713
6	1.403	13.860	0.995
7	1.868	18.453	1.325

frequency of the second main beam, $\omega_{6,2}$. The natural frequencies of the free-end main beams and the fixed-end sub-beam, normalized with respect to $\omega_a = \{(\omega_{5,1}^2 + \omega_{6,2}^2)/2\}^{1/2}$, are given in Table 1. Two window widths are considered: $\delta/\omega_0 = 0.25$ and 0.52. The smaller window contains mode 5 of the first main beam and mode 6 of the second main beam; the larger window also contains the first mode of the sub-beam. The central frequency value, ω_a, is used for ω_0 for the smaller window, and nine values of ω_0, which are distributed over the excitation frequency range, are used for the larger window. The translation responses are normalized with respect to the reponses of the first joint at $\omega = \omega_a$ for $\lambda = 3.0$. The results are plotted in Fig. 3 for $\lambda = 3.0$ and Fig. 4 for $\lambda = 10.0$.

The frequency window results show the essential characteristics of the response variation for both values of λ. Accuracy is enhanced by using more central frequencies ω_0 and a larger window.

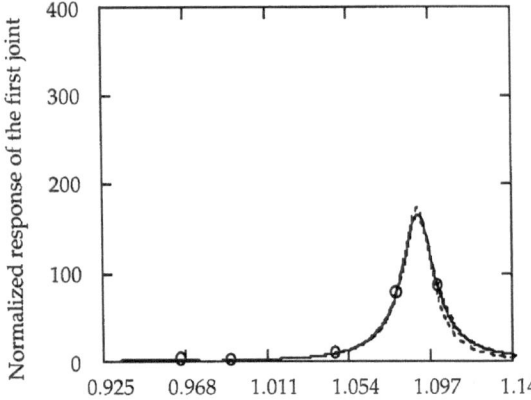

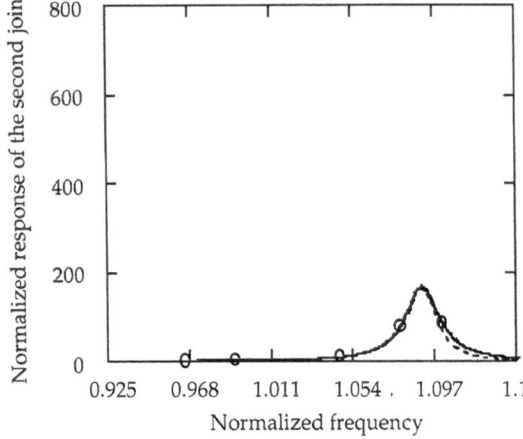

Fig. 4. Harmonic translation response for $\lambda = 10.0$ by 50-mode expansion (———) and frequency window method with $\delta/\omega_a = 0.25$ and $\omega_0 = \omega_a$, (- - - - -) and $\delta/\omega_a = 0.52$ and nine values of ω_0 (- -○- -)

Table 2. The normalized natural frequencies of main beams and sub-beam of the second example

Mode	Main beam 1 ω_{1i}/ω_1	Sub-beam ω_j/ω_1	Main beam 2 ω_{2i}/ω_1
1	0.088	1.000	0.063
2	0.243	2.757	0.172
3	0.477	5.404	0.338
4	0.788	8.933	0.559
5	1.177	13.345	0.835
6	1.644	18.638	1.166
7	2.189	24.814	1.552

The resonance peaks are located at frequencies which are shifted from the natural frequencies of mode 5 of the first main beam and mode 6 of the second main beam with free end conditions. This indicates strong coupling between the two modes of the main beams, which is caused by the sub-beam.

For the second example, the following nondimensional lengths are chosen: $l_2 = 1.188$ and $l_s = 0.21$. The harmonic response of the structure is determined for forcing frequencies in the neighborhood of the first natural frequency, ω_1, of the sub-beam. The natural frequencies of the free-end main beams and the fixed-end sub-beam, normalized with respect to ω_1, are shown in Table 2. One window width, $\delta/\omega_1 = 1.3$, is considered. This frequency window contains mode 1 of the sub-beam, modes 3, 4, 5, and 6 of the first main beam, and modes 4, 5, 6, and 7 of the second

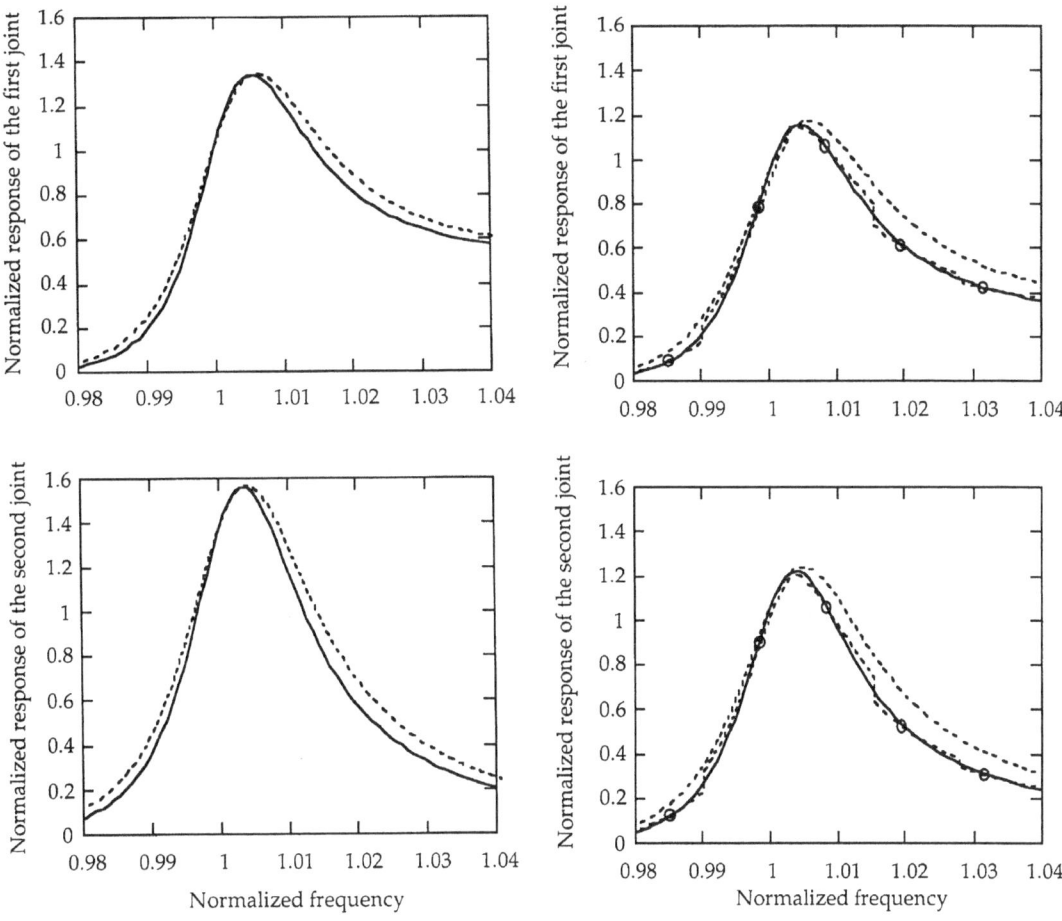

Fig. 5. Harmonic translation response for $\lambda = 3.0$ by 50-mode expansion (——) and frequency window method with $\delta/\omega_a = 1.3$ and $\omega_0 = \omega_1$ (- - - -)

Fig. 6. Harmonic translation response for $\lambda = 10.0$ by 50-mode expansion (——) and frequency window method with $\delta/\omega_0 = 1.3$ and $\omega_0 = \omega_1$, (- - - -) and $\delta/\omega_0 = 1.3$ and five-values of ω_0 (- -O- -)

main beam. The central value of the frequency window, ω_0, is chosen to be ω_1. The translation responses are normalized with respect to the responses of the first joint at $\omega = \omega_1$ for $\lambda = 3.0$. The results are plotted in Fig. 5 for $\lambda = 3.0$ and Fig. 6 for $\lambda = 10.0$.

The figures show that the frequency window results are most accurate for ω in a small neighborhood of ω_0. Greater accuracy can be obtained by using five values of ω_0 which are evenly distributed over the frequency range of interest, $0.98 < \omega/\omega_1 < 1.04$. The results are shown by connected circles for $\lambda = 10.0$. The errors in the frequency window analysis can be estimated without knowledge of the 50-mode expansion results by examining the discontinuities in the response curves between successive values of ω_0. This information can be used to improve the accuracy of the method as described earlier in this section.

5 Conclusions

A Lagrange formulation is applied to several main structures connected by sub-structures. The resulting equations of motion are analyzed using a frequency window method. The method reduces the computational complexity of the problem by retaining dependence for terms associated with modes with natural frequencies within the window and making frequency-independent approximations for the remaining terms.

The examples show that strongly coupled structures need wider windows to accurately account for frequency coupling effects. In cases where the response over a wide frequency range is necessary, several windows can be used to subdivide the frequency range. Such windowing can be used to estimate the accuracy of the method without additional calculations.

Acknowledgement

This research was supported by the Office of Naval Research under Contract No. 88K-0514, Dr. P. B. Abraham, Scientific Officer. This support is gratefully acknowledged.

References

[1] Gladwell, G. M. L., Bishop, R. E. D.: Interior receptances of beams. J. Mech. Eng. Sci. **2**, 1 − 15 (1960).
[2] Nicholson, J. W., Bergman, L. A.: Free vibration of combined dynmical systems. J. Eng. Mech. **112**/1, 1 − 13 (1986).
[3] Kelkel, K.: Green's function and receptance for structures consisting of beams and plates. AIAA Journal **25**/11, 1482 − 1489 (1987).
[4] Jacquot, L., Soedel, W. H.: Vibrations of elastic surface systems carrying dynamic elements. J. Acoust. Soc. Am., 47, 1354 − 1358 (1970).
[5] Hallquist, J., Snyder, V. W.: Linear damped vibratory structures with arbitrary support conditions. ASME J. Appl. Mech. **40**, 312 − 313 (1973).
[6] Leissa, A. W.: On a curve veering aberration. J. Appl. Math. Phys. **25**, 99 − 111 (1974).
[7] Dowell, E. H.: On some general properties of combined dynamical systems. ASME J. Appl. Mech. **46**, 206 − 209 (1979).
[8] Sackman, J. L., Kelly, J. M.: Seismic analysis of internal equipment and components in structures. Engineering Structures 1, 179 − 190 (1979).
[9] Sackman, J. L., Der Kiureghian, A., Nour-Omid, B.: Dynamic analysis of light equipment in structures: Modal properties of the combined system. J. Eng. Mech. Div., ASCE **109**/1, 73 − 89 (1983).
[10] Igusa, T., Der Kiureghian, A.: Dynamic response of multiply supported secondary systems. J. Eng. Mech. **111**/1, 20 − 41 (1985).
[11] Igusa, T., Achenbach, J. D., Min, K.-W.: Resonance characteristics of connected subsystems: Theory and simple configurations. J. Sound Vibr. **146**/3, 407 − 421 (1991 a).
[12] Igusa, T., Achenbach, J. D., Min, K.-W.: Resonance characteristics of connected subsystems: General configurations. J. Sound Vibr. **146**/3, 423 − 437 (1991 b).
[13] Meirovitch, L., Kwak, M.: On the modeling of flexible multi-body systems by the Rayleigh-Ritz method. Proceedings, AIAA dynamics specialist conference, Long Beach, CA, April 5 − 6, pp. 517 − 526 (1990).

Authors' address: Prof. J. D. Achenbach, R. R. McCormick School of Engineering and Applied Science, Northwestern University, Evanston, IL 60208, U.S.A.

Acta Mechanica (1992) [Suppl] 3: 35–53
© Springer-Verlag 1992

Dynamic analysis of column and borehole problems in soils and rocks

D. E. Beskos and **I. Vgenopoulou**, Patras, Greece

Summary. The one-dimensional dynamic column and borehole problems of soil and rock mechanics are solved analytically-numerically. The poroelastic soil medium obeys the Vardoulakis-Beskos theory, while the poroelastic, fissured rock medium the Aifantis-Beskos theory. The quasi-static counterparts of these problems are analysed as special cases of the dynamic ones. Use of Laplace transform with respect to time reduces the column and borehole problems to ordinary differential equations with constant and variable coefficients, respectively. The transformed solution of these problems is obtained analytically for the column and by finite differences for the borehole problem and after a numerical Laplace transform inversion, produces the time domain response. Viscoelastic material behavior of the solid skeleton is easily treated in the transformed domain with the aid of the correspondence principle. Both a suddenly applied and a harmonically varying with time load are considered. Numerical results are presented in order to assess the significance of various dynamic and material parameters on the response.

1 Introduction

The study of the dynamic behavior of fluid-saturated porous soils and rocks is of great importance in geotechnical engineering and geophysics. Biot [1] was the first to develop a three-dimensional theory of wave propagation in fluid-saturated porous elastic solids. The theory of mixtures (e.g., Bowen [2]) provided a rational way for constructing more general models to describe the quasi-static or dynamic behavior of fluid-filled porous media. An extensive review of wave propagation in porous media has been reported very recently by Corapcioglu [3]. The present work deals with the solution of the one-dimensional column and borehole problems of soil and rock mechanics and actually represents a review of the work very recently reported in Vgenopoulou and Beskos [4–6].

A brief review of some of the most important mixture-type models can be found in Vardoulakis and Beskos [7] and the references mentioned therein. One of these models due to Vardoulakis and Beskos [7] describes the dynamic behavior of nearly and fully saturated poroelastic soils, in contrast to most of the other models and the model of Biot [1], which are restricted to fully saturated conditions. This model has been constructed by using the theory of mixtures as a general framework and employing refined constitutive assumptions mainly due to Aifantis [8] and Verruijt [9, 10]. The first part of this work deals with the analytical-numerical solution of the one-dimensional dynamic column and borehole problems of soil mechanics formulated on the basis of the Vardoulakis-Beskos [7] model. In connection with the dynamic column problem of soil mechanics based on Biot type modelling one can mention here the one-dimensional analytical-numerical solutions of Garg et al. [11], Bowen and Lockett [12], Zienkiewicz et al. [13], Simon et al. [14], Vardoulakis and Beskos [7] and Hiremath et al. [15], the two-dimensional finite element solutions of Ghaboussi and Wilson [16], Prevost [17, 18], Zienkiewicz and Shiomi [19], Simon et al. [20, 21] and Hiremath et al. [15] and the two-dimensional boundary element solution of Cheng et al. [22]. In connection with the dynamic borehole problem of soil mechanics on the other hand, one can only mention the analytical-numerical work of Chang et al. [23] and Liu [24] on harmonic propagation analysis.

In the first part of this paper, as it was mentioned before, the one-dimensional column and borehole problems of soil mechanics formulated on the basis of the Vardoulakis-Beskos [7] theory are solved analytically-numerically. Use of Laplace transform with respect to time reduces the column and borehole problems to ordinary differential equations with constant and variable coefficients, respectively. The transformed solution of these problems is obtained analytically for the column and by finite differences for the borehole problem and after a numerical Laplace transform inversion produces the time domain response. Dynamic and quasi-static conditions are treated in a unified way in order to assess the inertial effects, while both harmonic with time and transient disturbances are considered in order to assess the loading type effects. In connection with the column problem, the present approach is similar to the ones in [7, 11, 15] but more accurate. In connection with the borehole problem, both harmonic and transient responses are obtained in this work for the first time by a method, which is also used for the first time for this type of problem. Furthermore, this work employs the more refined poroelastic theory of Vardoulakis-Beskos [7], which not only includes compressibilities of the constituents, but the effect of the degree of saturation as well.

Rock media are usually characterized by fissures separating them into porous blocks. In fissured rocks the permeability of the fissures is much higher than the permeability of the pores, while the porosity of the porous blocks is much larger than the porosity of the fissures. This clearly indicates that more refined models than Biot's [1] are needed for describing the quasi-static and dynamic behavior of these media. Beskos [25] has discussed this problem in a recent paper, which also contains an extensive literature review on the subject and describes the construction of the governing equations of motion for fully saturated, fissured, poroelastic rock media on the basis of Aifantis' [8] model of double porosity and the basic principles of the continuum theory of mixtures. In two previous papers Beskos et al. [26, 27] studied analytically-numerically the propagation of plane harmonic body and Rayleigh waves in such rock media.

In the second part of this work the one-dimensional dynamic column and borehole problems of rock mechanics, formulated on the basis of the rock model of Aifantis-Beskos [8, 25], are solved analytically-numerically. Both dynamic and quasi-static conditions are considered in order to assess the inertial effect and both harmonic with time and transient loads are applied in order to assess the effect of the load type on the response. The method is exactly the same as the one previously described for soils. The column and borehole problems of double porosity rock mechanics are solved here for the first time under dynamic conditions. The column problem under quasi-static conditions has been solved by Wilson and Aifantis [28] and Beskos and Aifantis [29] analytically and by Khaled et al. [30] by two-dimensional finite elements on the basis of Aifantis' [8] rock model and by Ohnishi et al. [31] and Valliappan and Khalili-Naghadeh [32] by two-dimensional finite elements on the basis of their own double porosity rock models. The borehole problem has also been solved analytically under quasi-static conditions and on the basis of Aifantis' [5] rock model by Wilson and Aifantis [28] and Beskos and Aifantis [29]. However, their final solution expressions require the evaluation of complicated integrals, which was not attempted in those papers.

The effect of the viscoelastic material behavior of the solid skeleton in soils and rocks, modelled after Vardoulakis and Beskos [7] and Aifantis and Beskos [8, 25], respectively, on the response associated with the one-dimensional column problem is also studied in this work. Biot [33] was the first to include in his model viscoelastic behavior of the soil skeleton and study the solution of the column consolidation problem. Tabaddor and Little [34], Podstrigach and Pavlina [35] and Predeleanu [36] also considered fluid flow in viscoelastic porous media under quasi-static conditions, while Dziecielak [37] wave propagation in fluid-saturated porovisco-elastic media. The subject of fluid flow through poroviscoelastic media under quasi-static conditions was treated in the most general and complete manner by Taylor and Aifantis [38]. In this work the formalism and ideas of Taylor and Aifantis [38] are used in extending the dynamic poroelastic models of Vardoulakis-Beskos [7] for soil and Aifantis-Beskos [25] for rock to the dynamic poroviscoelastic case. Correspondence principles in the Laplace transform domain for both models are established and the corresponding dynamic column problems are solved analytically-numerically with the aid of their correspondence principles.

Numerical results are presented in this work for both the column and the borehole problems of soil and rock mechanics in order to assess the significance of various dynamic and material parameters on the response.

2 The dynamic soil column problem

Consider a soil column of height h, laterally confined and with a rigid and impervious bottom $z = h$, which is subjected at its top surface $z = 0$ to a suddenly applied compressive load of magnitude σ_0. This load is applied through a porous slab so that water can escape from the soil surface. The dynamic response of this column is to be determined herein. Vardoulakis and Beskos [7] have recently proposed a more refined model than Biot's [1] for the description of the dynamic behavior of fully and nearly saturated poroelastic soil media. This model reduces to the following system of equations for this one-dimensional problem:

$$(\lambda + 2\mu)\, u_{,zz} = -(1 - \gamma)\, bQ + \gamma \varrho_f \dot{Q} + (\bar{\varrho}_s + \gamma \bar{\varrho}_f)\, \ddot{u}, \tag{1}$$

$$-p_{,z} = bQ + \varrho_f \dot{Q} + \bar{\varrho}_f \ddot{u}, \tag{2}$$

$$-Q_{,z} = \alpha_1 \dot{u}_{,z} + \alpha_2 \dot{p}. \tag{3}$$

In the above, commas indicate differentiation with respect to z along the column, dots differentiation with respect to time t, u is the solid phase displacement, p is the fluid pore pressure, $\bar{\varrho}_s$ and $\bar{\varrho}_f$ are the relative solid and fluid densities given in terms of the actual ones by the relations

$$\bar{\varrho}_s = (1 - n)\, \varrho_s, \qquad \bar{\varrho}_f = n \varrho_f \tag{4}$$

where n is the porosity, λ and μ are the Lamé elastic constants, γ, α_1, and α_2 are material constants given by

$$\gamma = \beta_s/\beta, \qquad \beta = 1/(\lambda + 2\mu), \tag{5}$$

$$\alpha_1 = n + (1 - n)\, [1 - (\beta_p/\beta)], \tag{6}$$

$$\alpha_2 = n\beta_f{}^* + (1 - n)\, [1 - (\beta_p/\beta)]\, \beta_s, \tag{7}$$

$$\beta_f{}^* = \beta_f\{1 + [(1 - S)/p_g\beta_f]\} \tag{8}$$

where β is the bulk compressibility of the solid skeleton, β_s is the compressibility of the solid particle material, β_p is the solid particle compressibility due to concentrated forces at the contact point among particles, β_f is the fluid compressibility, β_f^* is the compressibility of the fluid-air mixture near saturation, S is the degree of saturation ($0.95 \leq S < 1$ for near saturation and $S = 1$ for full saturation) and $p_g = \varrho_f gh$ is a reference hydrostatic pressure with g being the acceleration of gravity and h the depth below ground surface, and b and Q are the fluid seepage force factor and the relative specific discharge, respectively, given by

$$b = v_f/K, \tag{9}$$

$$Q = n(\dot{v} - \dot{u}) \tag{10}$$

where v_f is the dynamic viscosity of the fluid, K is the Muskat permeability and v is the fluid displacement.

The boundary and initial conditions of the problem read (Vgenopoulou and Beskos [4])

$$p(0, t) = 0, \qquad \sigma_z(0, t) = -\sigma_0, \tag{11}$$

$$u(h, t) = 0, \qquad Q(h, t) = 0 \quad \text{or} \quad p_{,z}(h, t) = 0 \tag{12}$$

and

$$Q(z, 0) = \dot{u}(z, 0) = 0, \qquad u(z, 0) = (z - h)\, u_0, \tag{13}$$

$$p(z, 0) = -(\alpha_1/\alpha_2)\, u_0, \tag{14}$$

$$u_0 = -\sigma_0 \alpha_2/[\alpha_2(\lambda + 2\mu) + (1 - \gamma)\, \alpha_1] \tag{15}$$

respectively.

Application of Laplace transform with respect to time to Eqs. (1)–(3) under the initial conditions (13)–(15) and after elimination of $\bar{Q}$ yields

$$A_{11}\bar{u}_{,z} + A_{12}\bar{p}_{,zz} + A_{13}\bar{p} = F_1,$$
$$\tag{16}$$
$$A_{21}\bar{u}_{,zz} + A_{22}\bar{u} + A_{23}\bar{p}_{,z} = F_2(z - h)$$

where overbars denote Laplace transformed quantities and

$$A_{11} = [\bar{\varrho}_f s^2/(b + \varrho_f s)] - \alpha_1 s,$$

$$A_{12} = 1/(b + \varrho_f s), \qquad A_{13} = -\alpha_2 s,$$

$$F_1 = \bar{\varrho}_f s u_0/(b + \varrho_f s),$$
$$\tag{17}$$
$$A_{21} = \lambda + 2\mu, \qquad A_{23} = [\gamma \varrho_f s + (\gamma - 1)\, b]/(b + \varrho_f s),$$

$$A_{22} = \{\bar{\varrho}_f s^2[\gamma \varrho_f s + (\gamma - 1)\, b]/(b + \varrho_f s)\} - (\bar{\varrho}_s + \gamma \bar{\varrho}_f)\, s^2,$$

$$F_2 = \{[\gamma \varrho_f s + (\gamma - 1)\, b]\, \bar{\varrho}_f/(b + \varrho_f s)\}\, s u_0 + (\bar{\varrho}_s + \gamma \bar{\varrho}_f)\, s u_0$$

with s being the Laplace transform parameter.

Assuming a solution for the homogeneous part of the system (16) of the form

$$\bar{u}_h = C_1 e^{\xi z}, \qquad \bar{p}_h = C_2 e^{\xi z} \tag{18}$$

the condition for nontrivial solutions is

$$\xi^4 + B\xi^2 + C = 0 \tag{19}$$

where

$$B = (A_{21}A_{13} + A_{12}A_{22} - A_{11}A_{23})/A_{12}A_{21},$$
$$C = A_{13}A_{22}/A_{12}A_{21}. \tag{20}$$

Thus the homogeneous solution has the form

$$\bar{u}_h = \sum_{j=1}^{4} C_{1j}e^{\xi_j z}, \quad \bar{p}_h = \sum_{j=1}^{4} C_{2j}e^{\xi_j z} \tag{21}$$

where ξ_j ($j = 1, 2, 3, 4$) are the four complex roots of (19) and

$$C_{1j} = -[(A_{12}\xi_j^2 + A_{13})/A_{11}\xi_j] C_{2j}. \tag{22}$$

Assuming now a particular solution of (16) of the form

$$\bar{u}_p = B_1(z - h), \quad \bar{p}_p = B_2 \tag{23}$$

one can easily obtain

$$B_1 = F_2/A_{22},$$
$$B_2 = (A_{22}F_1 - A_{11}F_2)/A_{13}A_{22}. \tag{24}$$

Thus the complete solution of (16) has the form

$$\bar{u}(z, s) = \bar{u}_h + \bar{u}_p, \quad \bar{p}(z, s) = \bar{p}_h + \bar{p}_p \tag{25}$$

where its homogeneous and particular parts are given by (21) and (23), respectively, with the four constants C_{2j} ($j = 1, 2, 3, 4$) being computed with the aid of the four boundary conditions (11) and (12) in the transformed form

$$\bar{p}(0, s) = 0, \quad \bar{u}_{,z}(0, s) = -\sigma_0/(\lambda + 2\mu) s,$$
$$\bar{u}(h, s) = 0, \quad \bar{p}_{,z}(h, s) = 0. \tag{26}$$

The time domain response of the soil column is finally obtained by a numerical inversion of the Laplace transformed solution (25), which employs the algorithm of Durbin [39]. This algorithm has been found in Narayanan and Beskos [40] to be the most accurate one for dynamic problems. The present method gives almost identical results with the analytical method of Bowen and Lockett [12] as it is explained in [4]. However, the present method is simpler and more general than the one in [12].

The corresponding quasi-static (consolidation) column problem can be analysed as a special case of the dynamic problem. Thus for $\ddot{u} = \dot{Q} = 0$ and after eliminating Q and applying Laplace transform with respect to time, Eqs. (1)−(3) become

$$(\lambda + 2\mu) \bar{u}_{,zz} - (1 - \gamma) \bar{p}_{,z} = 0,$$
$$\bar{p}_{,zz} - b\alpha_1 s\bar{u}_{,z} - b\alpha_2 s\bar{p} = 0 \tag{27}$$

with a solution of the form

$$\bar{p} = C_1 e^{az} + C_2 e^{-az} + C_3,$$

$$\bar{u} = \frac{(1-\gamma)}{(\lambda + 2\mu)} \left(\frac{C_1}{a} e^{az} - \frac{C_2}{a} e^{-az} + C_3 z + C_4 \right), \tag{28}$$

$$a^2 = bs\{\alpha_2 + [\alpha_1(1-\gamma)/(\lambda + 2\mu)]\}.$$

The four constants of integration C_j ($j = 1, 2, 3, 4$) are computed with the aid of boundary conditions (26) and the time domain solution is finally obtained by a numerical Laplace transform inversion of (28) as described in the dynamic case. The present method gives almost identical results with the analytical method of Biot [41] as it is explained in [4].

In addition to the suddenly applied load case discussed above, the case of a harmonically varying with time surface load $\sigma_0 \cos \omega t$, where ω is the circular operational frequency, can also be treated by the method just described under both dynamic and quasi-static conditions.

3 The dynamic soil borehole problem

Consider a vertical borehole in soil of circular cross-section with radius r_0. The response of the surrounding soil to a sudden pressurization of magnitude p_0 of the borehole is to be determined. Under conditions of plane strain and axial symmetry, the above problem becomes one-dimensional with the radial coordinate r being the only spatial coordinate. The Vardou-lakis-Beskos [7] soil model reduces to the following system of equations for this one-dimensional problem:

$$(\lambda + 2\mu) [u_{,rr} + (1/r) u_{,r} - (1/r^2) u] = -(1-\gamma) bQ + \gamma \varrho_f \dot{Q} + (\bar{\varrho}_s + \gamma \bar{\varrho}_f) \ddot{u}, \tag{29}$$

$$-p_{,r} = bQ + \varrho_f \dot{Q} + \bar{\varrho}_f \ddot{u}, \tag{30}$$

$$-Q_{,r} - (1/r) Q = \alpha_1[\dot{u}_{,r} + (1/r) \dot{u}] + \alpha_2 \dot{p}. \tag{31}$$

The boundary and initial conditions of the problem read (Vgenopoulou and Beskos [4])

$$p(r_0, t) = p_0, \qquad \sigma_r(r_0, t) = -p_0, \tag{32}$$

$$p(\infty, t) = 0, \qquad \sigma_r(\infty, t) = 0 \tag{33}$$

and

$$Q(r, 0) = \dot{u}(r, 0) = 0, \tag{34}$$

$$p(r, 0) = 0, \qquad u(r, 0) = 0, \tag{35}$$

respectively.

Application of Laplace transform with respect to time to Eqs. (29)−(31) under the initial conditions (34) and (35) and after elimination of $\bar{Q}$ yields

$$B_{11} \left(\bar{u}_{,rr} + \frac{1}{r} \bar{u}_{,r} \right) + B_{12} \bar{u} + B_{13} \bar{p}_{,r} = 0,$$

$$\tag{36}$$

$$B_{21} \left(\bar{p}_{,rr} + \frac{1}{r} \bar{p}_{,r} \right) + B_{22} \bar{p} + B_{23} \bar{u}_{,r} + B_{24} \bar{u} = 0$$

where

$$B_{11} = (\lambda + 2\mu)(b + \varrho_f s),$$

$$B_{12} = \bar{\varrho}_f s^2 [\gamma \varrho_f s - (1 - \gamma) b] - (b + \varrho_f s)(\bar{\varrho}_s + \gamma \bar{\varrho}_f) s^2 - (\lambda + 2\mu)(b + \varrho_f s)(1/r^2),$$

$$B_{13} = \gamma \varrho_f s - (1 - \gamma) b, \qquad B_{21} = 1, \tag{37}$$

$$B_{22} = \alpha_2 s(b + \varrho_f s), \qquad B_{23} = \bar{\varrho}_f s^2 - \alpha_1 s(b + \varrho_f s),$$

$$B_{24} = (1/r)[\varrho_f s^2 - \alpha_1 s(b + \varrho_f s)].$$

The system of Eq. (36) involves two coupled linear ordinary differential equations with variable coefficients. Its solution is obtained numerically by employing the central difference method (von Rosenberg [42]) in conjunction with the boundary conditions (32) and (33) as described in Vgenopoulou and Beskos [4]. The r axis is divided into N equal segments of length Δr and Eq. (36) creates $2N$ linear equations with $2N$ unknowns ($\bar{u}_i$, $\bar{p}_i$, $i = 1, 2, ..., N$). The number N is large, since r ranges from r_0 to infinity. Here a value of $N = 1\,200$ for $\Delta r = 0.6$ was used corresponding to $r_0 \leqq r \leqq 720 r_0$. Finally a numerical inversion of the transformed solution with the aid of the algorithm of Durbin [39] can provide the time domain response. The corresponding quasi-static (consolidation) borehole problem can be analysed as a special case of the dynamic problem. Thus for $\ddot{u} = \dot{Q} = 0$ and after eliminating Q and applying Laplace transform with respect to time, Eqs. (29)–(31) become

$$(\lambda + 2\mu)\left(\bar{u}_{,rr} + \frac{1}{r}\bar{u}_{,r} - \frac{1}{r^2}\bar{u}\right) - (1 - \gamma)\bar{p}_{,r} = 0,$$

$$\tag{38}$$

$$\bar{p}_{,rr} + \frac{1}{r}\bar{p}_{,r} - \alpha_2 bs\bar{p} - \alpha_1 bs\left(\bar{u}_{,r} + \frac{1}{r}\bar{u}\right) = 0.$$

The above equation is solved numerically by finite differences and the transformed solution is inverted numerically to produce the time domain response in exactly the same way as in the dynamic case. The present method gives almost identical results with the analytical-numerical method of Vgenopoulou and Beskos [4] for the soil borehole problem under quasi-static conditions.

In addition to the suddenly applied pressure case discussed above, the case of a harmonically varying with time pressure $p_0 \cos \omega t$ inside the borehole can also be treated by the method just described under both dynamic and quasi-static conditions.

4 The dynamic rock column problem

Consider the column problem with geometry, kind of loading and boundary conditions those of Sect. 2. The column material is assumed to be a fissured rock obeying the Aifantis-Beskos [8, 25] rock theory. A description of the dynamic behavior of this fully saturated, fissured, poroelastic rock model can be found in Beskos [25]. This model is characterized by two degrees of porosity — one due to the fissures, separating the rock into blocks and the other due to the pores of those blocks — and reduces to the following system of equations for the one-dimensional dynamic rock column problem:

$$(\lambda + 2\mu) u_{,zz} = \nu_1 Q_1 - \nu_2 Q_2 + \bar{\varrho}_s \ddot{u}, \tag{39}$$

$$-\beta_\alpha p_{\alpha,z} = \nu_\alpha Q_\alpha + \varrho_f \dot{Q}_\alpha + \bar{\varrho}_\alpha \ddot{u}, \tag{40}$$

$$(\beta_\alpha + n_\alpha) \dot{u}_{,z} + (\gamma_\alpha + \beta_f n_\alpha) \dot{p}_\alpha + Q_{\alpha,z} = -(-1)^\alpha \varkappa(p_2 - p_1). \tag{41}$$

In the above, α receives the values 1 (for fissures) and 2 (for pores) and does not imply summation when it is repeated, the coefficients β_α express the deformability of the solid as this affects the flow in the fissures ($\alpha = 1$) and the pores ($\alpha = 2$), the coefficients γ_α measure the compressibility of the fissures ($\alpha = 1$) and the pores ($\alpha = 2$), the coefficient $\varkappa$ measures the transfer of fluid from the pores to the fissures and the seepage force factors v_α and the relative specific discharges Q_α in the fissures ($\alpha = 1$) and the pores ($\alpha = 2$) are given by

$$v_\alpha = \beta_\alpha v_f / K_\alpha, \tag{42}$$

$$Q_\alpha = n_\alpha(\dot{u}_\alpha - \dot{u}), \tag{43}$$

with v_f being the dynamic viscosity of the fluid, K_α the Muskat permeabilities of the fissures ($\alpha = 1$) and the pores ($\alpha = 2$), and u_α the displacements of the fluid in the fissures ($\alpha = 1$) and the pores ($\alpha = 2$). The remaining quantities have been defined before with the exception of $\bar{\varrho}_s$ and $\bar{\varrho}_\alpha$, which are now defined by

$$\bar{\varrho}_s = \varrho_s[1 - (n_1 + n_2)], \qquad \bar{\varrho}_\alpha = n_\alpha \varrho_f. \tag{44}$$

The boundary and initial conditions of this column problem read (Vgenopoulou and Beskos [5])

$$p_\alpha(0, t) = 0, \qquad \sigma_z(0, t) = -\sigma_0, \tag{45}$$

$$u(h, t) = 0, \qquad Q_\alpha(h, t) = 0 \quad \text{or} \quad p_{\alpha,z}(h, t) = 0, \tag{46}$$

and

$$Q_\alpha(z, 0) = \dot{u}(z, 0) = 0, \qquad u(z, 0) = (z - h)\, u_0, \tag{47}$$

$$p_\alpha(z, 0) = [(\beta_\alpha + n_\alpha)/(\gamma_\alpha + \beta_f n_\alpha)]\, u_0, \tag{48}$$

$$u_0 = -\sigma_0 \left/ \left[(\lambda + 2\mu) + \beta_1 \left(\frac{\beta_1 + n_1}{\gamma_1 + \beta_f n_1} \right) + \beta_2 \left(\frac{\beta_2 + n_2}{\gamma_2 + \beta_f n_2} \right) \right], \right. \tag{49}$$

respectively.

The solution of the above problem follows exactly the solution procedure of Sect. 2 used for the corresponding problem with single porosity. The only difference is now that one has a system of three instead of two ordinary differential equations in the Laplace transformed domain to solve analytically. For more details one can consult Vgenopoulou and Beskos [5]. The special cases of quasi-static conditions and harmonically varying with time loading are also treated in the same manner as in Sect. 2. For more details reference [5] can again be consulted.

5 The dynamic rock borehole problem

Consider the borehole problem with geometry, kind of loading and boundary conditions those of Sect. 3. The material of the medium surrounding the borehole is assumed to be a fissured rock obeying the Aifantis-Beskos [8, 25] rock theory. The governing equations of motion of this fully saturated, fissured, poroelastic rock medium developed in Beskos [25] are reduced to the following system of equations for the one-dimensional dynamic borehole problem:

$$(\lambda + 2\mu)\, [u_{,rr} + (1/r)\, u_{,r} - (1/r^2)\, u] = -v_1 Q_1 - v_2 Q_2 + \bar{\varrho}_s \ddot{u}, \tag{50}$$

$$-\beta_\alpha p_{\alpha,r} = v_\alpha Q_\alpha + \varrho_f \dot{Q}_\alpha + \bar{\varrho}_\alpha \ddot{u}, \tag{51}$$

$$(\beta_\alpha + n_\alpha)\, [\dot{u}_{,r} + (1/r)\, \dot{u}] + (\gamma_\alpha + \beta_f n_\alpha)\, \dot{p}_\alpha + Q_{\alpha,r} + (1/r)\, Q_\alpha = -(-1)^\alpha\, \varkappa(p_2 - p_1). \tag{52}$$

The boundary and initial conditions of the above problem read (Vgenopoulou and Beskos [5])

$$p_\alpha(r_0, t) = p_0, \qquad \sigma_r(r_0, t) = -p_0, \tag{53}$$

$$p_\alpha(\infty, t) = 0, \qquad \sigma_r(\infty, t) = 0, \tag{54}$$

and

$$Q_\alpha(r, 0) = \dot{u}(r, 0) = 0, \tag{55}$$

$$p_\alpha(r, 0) = 0, \qquad u(r, 0) = 0, \tag{56}$$

respectively.

The solution of the above problem follows exactly the solution procedure of Sect. 3 used for the corresponding problem with single porosity. The only difference is now that one has a system of three instead of two ordinary differential equations in the Laplace transformed domain to solve numerically. For more details one can consult Vgenopoulou and Beskos [5]. The special cases of quasi-static conditions and harmonically varying with time loading are also treated in the same manner as in Sect. 3. For more details reference [5] can again be consulted.

6 The effect of viscoelasticity

The Vardoulakis-Beskos [7] model for the dynamics of nearly and fully saturated poroelastic soils and the Aifantis-Beskos [8, 25] model for the dynamics of fully saturated, fissured, poroelastic rocks are here extended to include viscoelastic material behavior. Linear hereditary isotropic viscoelasticity of the relaxation type for the solid phase is considered. Thus one has (Taylor and Aifantis [38])

$$\sigma_{ij}(x, t) = G_{ijkl}(t) * \varepsilon_{kl}(x, t) \tag{57}$$

where σ_{ij} and ε_{kl} are the stress and strain tensors, respectively, x represents a point in the body, $G_{ijkl}(t)$ is the relaxation response function and the symbol $*$ is defined through the Boltzmann's operator $G*$ acting on a vector function $F(x, t)$ by the relation

$$G(t) * F(x, t) = G(0)\, F(x, t) + \int_0^\infty F(x, t - s)\, [dG(s)/ds]\, ds. \tag{58}$$

Following the work of Taylor and Aifantis [38] one can easily extend the Vardoulakis-Beskos [7] poroelastic soil model or the Aifantis-Beskos [8, 25] poroelastic rock model to a porovisco-elastic one. Starting with the constitutive relations of the poroelastic model and assuming that there is a linear dependence of the various quantities on the histories of the solid strain, the relative solid-fluid velocity and the fluid pressure, one can obtain the time-dependent constitutive equations of the poroviscoelastic model. These are substituted into the field equations to obtain the governing equations of motion of the poroviscoelastic model. These equations of the poroviscoelastic model are Laplace transformed with respect to time and compared with the Laplace transformed governing equations of motion of the poroelastic model. This comparison reveals that one can go from the poroelastic to poroviscoelastic case in the transformed domain if the elastic material constants of the former case are replaced by the transformed viscoelastic functions of the latter case as follows:

$$\lambda \to s\bar{\lambda}^*, \quad \mu \to s\bar{\mu}^*, \quad \gamma \to s\bar{\gamma}^*, \quad b \to s\bar{b}^*,$$

$$\alpha_1 \to s\bar{\alpha}_1^*, \quad \alpha_2 \to s\bar{\alpha}_2^*, \quad (\beta_p/\beta) \to s\overline{(\beta_p/\beta)}^*, \quad \beta_f \to s\bar{\beta}_f^*, \quad \beta_s \to s\bar{\beta}_s^* \tag{59}$$

for the soil model and

$$\lambda \to s\bar{\lambda}^*, \quad \mu \to s\bar{\mu}^*, \quad \beta_\alpha \to s\bar{\beta}_\alpha^*,$$

$$\nu_\alpha \to s\bar{\nu}_\alpha^*, \quad \beta_f \to s\bar{\beta}_f^*, \quad \gamma_\alpha \to s\bar{\gamma}_\alpha^* \tag{60}$$

for the rock model, where s is the Laplace transformed parameter. Eqs. (59) and (60) constitute the correspondence principles of viscoelasticity for the soil and rock models considered here. For more details on this subject one can consult Vgenopoulou and Beskos [6].

To the authors' best knowledge there are no experimentally obtained expressions in the literature for the various viscoelastic response functions of the relaxation type employed here. Thus it was assumed that these functions have a time variation of the form

$$G^*(t) = G_0 + G_1 \exp(-t/\tau_0) \tag{61}$$

where G^* represents any one of the viscoelastic response functions of Eqs. (59) and (60) and

$$G_0 = G/5, \quad G_0 + G_1 = G$$

with G being the corresponding elastic constant and where τ_0 is a time constant taken here to be 458.5 secs.

The above viscoelastic functions were used in conjunction with the corresponding elastic material constants of the aforementioned soil and rock models in the soil and rock column problems under both quasi-static and dynamic conditions. Following the solution procedure already described in Sects. 2 and 4 in conjunction with the correspondence principles (59) and (60) for soils and rocks, respectively, one can easily obtain the response in the time domain. More details can be found in reference [6].

7 Numerical results and discussion

Numerical results were obtained on the basis of soil and rock data described in Tables 1 and 2, respectively. Figures 1 and 2 show the displacement and pressure histories at soil column height of $h/2 = 10$ m for two degrees of saturation S for the case of the suddenly applied load under both

Table 1. Material properties of granular soil

Porosity	$n = 0.45$
Soil Poisson's ratio	$v = 0.25$
Soil compressibility	$1/\beta = 341$ MPa
Water compressibility	$1/\beta_f = 2.14$ GPa
Relative compressibility	$\gamma = 0.0159$
Solid density	$\varrho_s = 2\,660$ kg/m³
Water density	$\varrho_f = 1\,000$ kg/m³
Soil permeability	$K = 4.008 \times 10^{-10}$ m²
Water dynamic viscosity	$v_f = 1.002 \times 10^{-3}$ Ns/m²

Table 2. Numerical values of coefficients for water-fully saturated, poroelastic, fissured sandstone

Lamé constant	λ	2.76×10^9	N/m^2
Lamé constant	μ	2.76×10^9	N/m^2
Porosity	n_α	$n_1 = 0.0236$	—
		$n_2 = 0.2364$	—
Porosity change due	β_α	$\beta_1 = 0.20$	—
to dilatation		$\beta_2 = 0.70$	—
Solid density	ϱ_s	2.66×10^3	kg/m^3
Water density	ϱ_f	1.00×10^3	kg/m^3
Water compressibility	β_f	4.67×10^{-10}	$1/N \cdot m^2$
Compressibility of	γ_α	$\gamma_1 = 4.34 \times 10^{-9}$	$1/N \cdot m^2$
fissures and pores		$\gamma_2 = 4.24 \times 10^{-9}$	$1/N \cdot m^2$
Water dynamic viscosity	v_f	1.00×10^{-3}	$N \cdot s/m^2$
Muskat permeability	K_α	$K_1 = 7.92 \times 10^{-10}$	m^2
		$K_2 = 7.92 \times 10^{-13}$	m^2
Water transfer from fissures to pores	$\varkappa$	2.0×10^{-16}	$N/m^2 \cdot s$

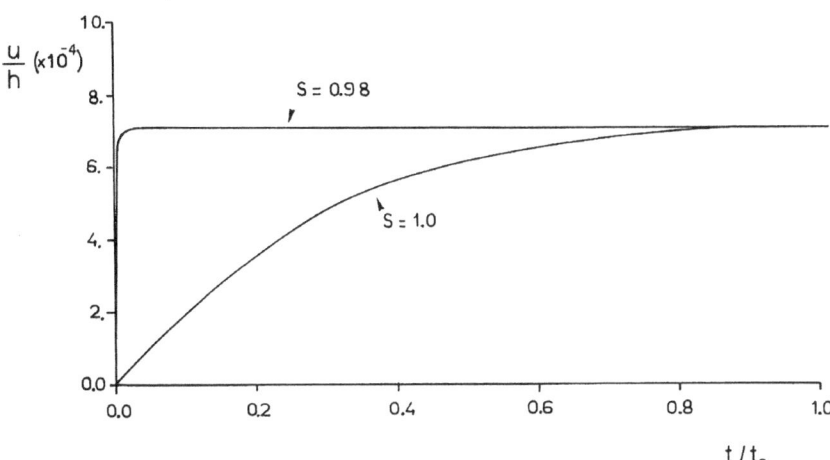

Fig. 1. Solid displacement history for nearly and fully saturated soil column under suddenly applied load

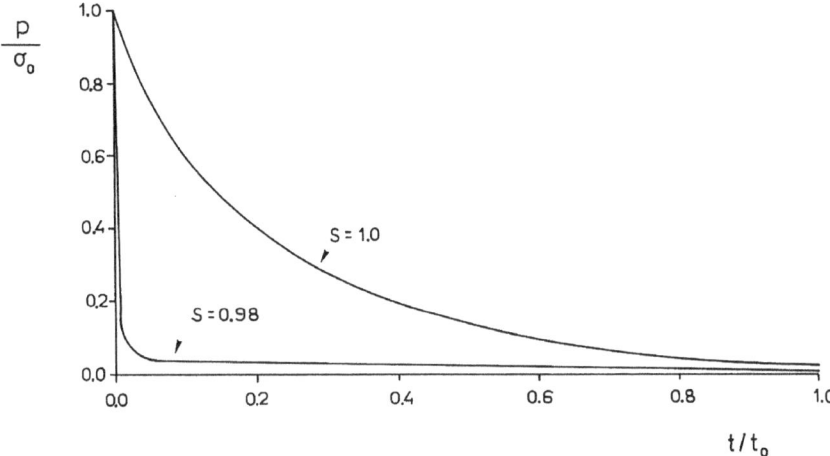

Fig. 2. Fluid pore pressure history for nearly and fully saturated soil column under suddenly applied load

dynamic and quasi-static conditions, which produce almost identical results. However, if the applied load is harmonically varying with time with $\omega = 50\,\text{sec}^{-1}$, then the dynamic response is higher than the quasi-static one as Fig. 3 clearly indicates for the case of the column displacement history at $h/2$. The normalizing factors σ_0 and t_0 are given by $\sigma_0 = 0.5\,\text{MPa}$ and $t_0 = h/(c_{rp}K)^{1/2}$, where $c_{rp}^2 = (\lambda + 2\mu)/(\bar{\varrho}_s + \gamma\bar{\varrho}_f)$. Figures 4 and 5

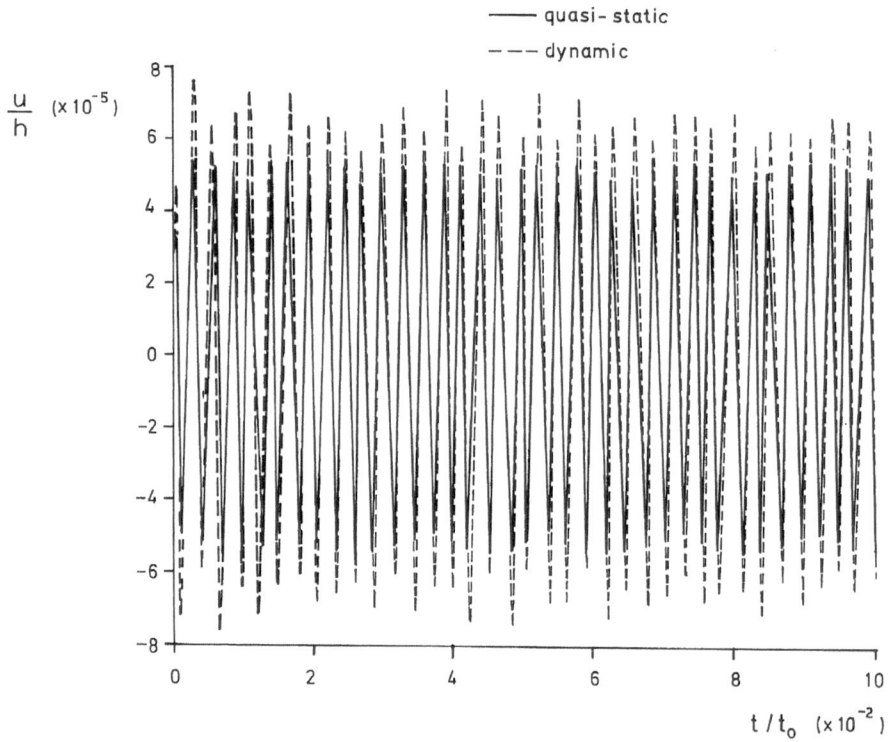

Fig. 3. Solid displacement history for fully saturated soil column under quasi-static and dynamic conditions subjected to a harmonic load

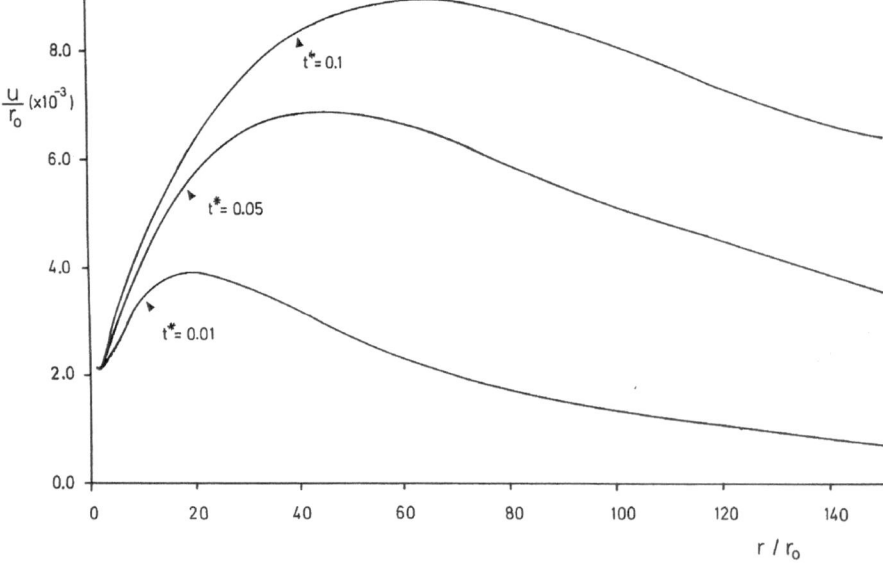

Fig. 4. Solid displacement versus radial distance at various times $t^* = t/t_0$ for fully saturated soil borehole problem under suddenly applied pressure

depict the displacement and pressure versus distance with $r_0 = 0.15$ m for various times $t^* = t/t_0$ for the soil borehole problem under dynamic and quasi-static conditions, producing almost identical results. However, if the applied pressure is harmonic with $\omega = 50 \sec^{-1}$, the dynamic response is higher than the quasi-static one as Fig. 6 clearly indicates.

Figures 7 and 8 portray the displacement and pore pressure histories at rock column height of $h/2 = 10$ m for various porosity ratios n_1/n ($n = n_1 + n_2 = 0.26$) and $K_1/K_2 = 1000$ for the case of the suddenly applied load under both dynamic and quasi-static conditions, which produce almost identical results. However, if the applied load is harmonic with

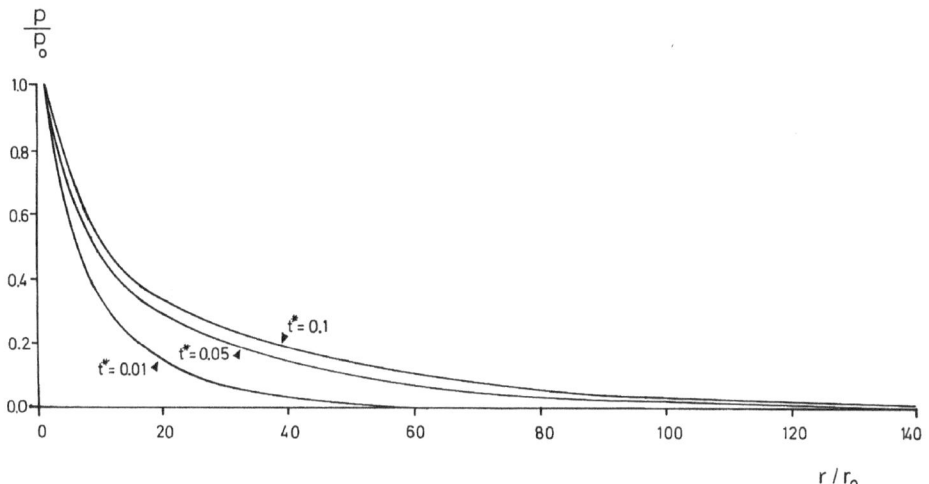

Fig. 5. Fluid pore pressure versus radial distance at various times $t^* = t/t_0$ for fully saturated soil borehole problem under suddenly applied pressure

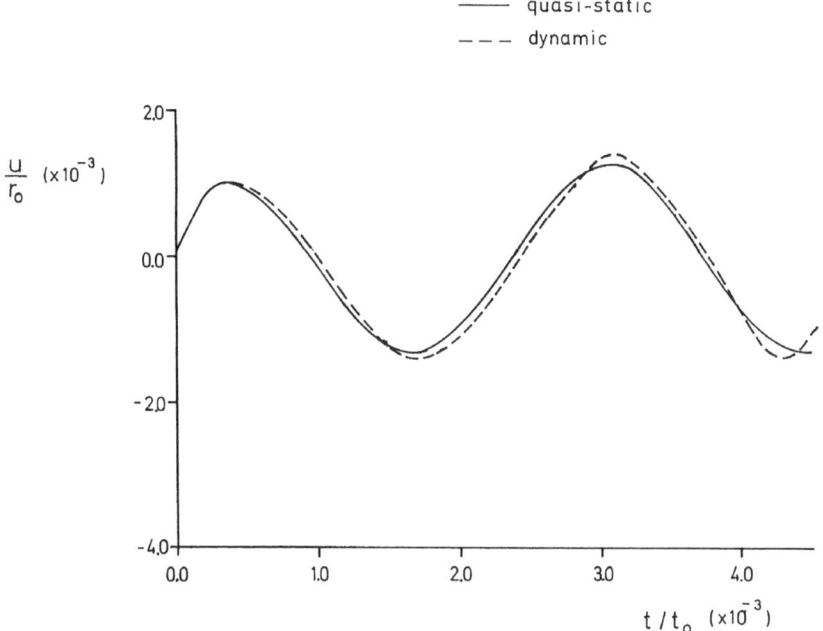

Fig. 6. Solid displacement history for fully saturated soil borehole under quasi-static and dynamic conditions subjected to a harmonic pressure

$\omega = 50 \sec^{-1}$, then the dynamic response is higher than the quasi-static one as Fig. 9 clearly indicates. The factor t_0 is now computed with $K = K_2$ and $c_{rp}^2 = (\lambda + 2\mu)/\bar{\varrho}_s$. Figures 10 and 11 depict the solid displacement and fissure pressure versus distance for various times $t^* = t/t_0$ for the rock borehole problem under dynamic and quasi-static conditions, producing almost identical results. However, if the applied pressure is harmonic, the dynamic response is higher than the quasi-static one as Fig. 12 clearly indicates.

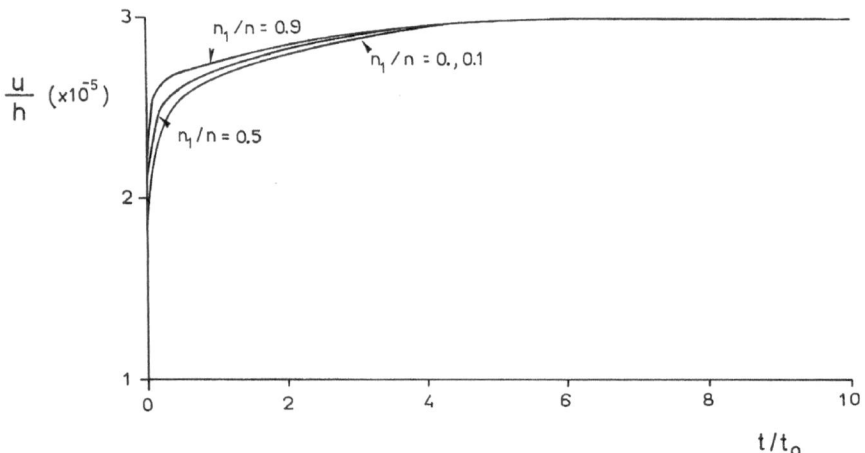

Fig. 7. Solid displacement history for various values of porosity ratio in rock column problem under suddenly applied load

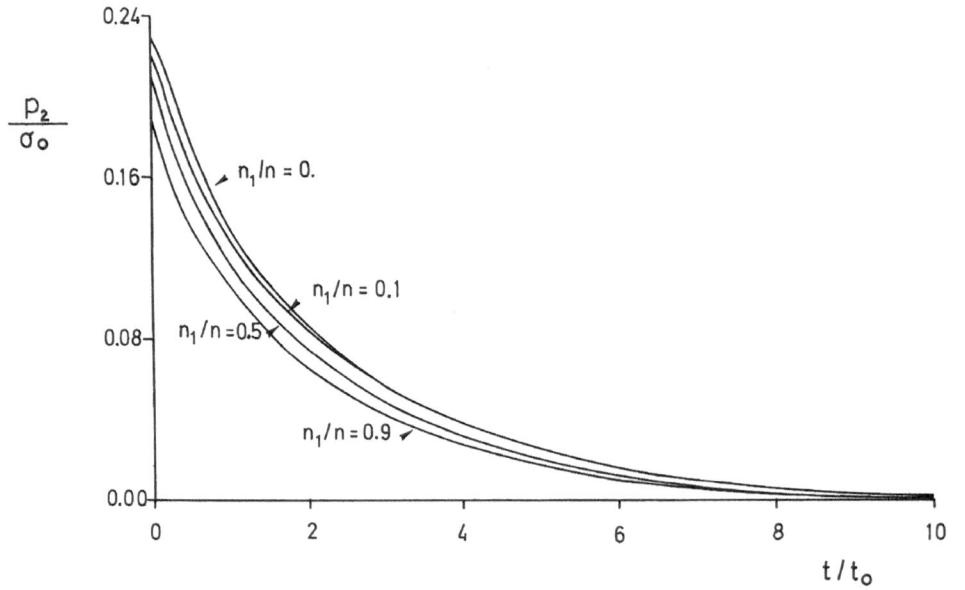

Fig. 8. Fluid pore pressure history for various values of porosity ratio in rock column problem under suddenly applied load

Figures 13 and 14 depict the histories of the solid displacement and pore pressure, respectively, at a column depth of $h/2 = 10$ m for $S = 1$ and 0.98 for both viscoelastic and elastic behavior of the soil skeleton. It is observed that the effect of viscoelasticity consists, in general, in increasing the elastic response, especially at late times, in qualitative agreement with Biot [33]. Analogous results hold true for the case of the rock column problem [6].

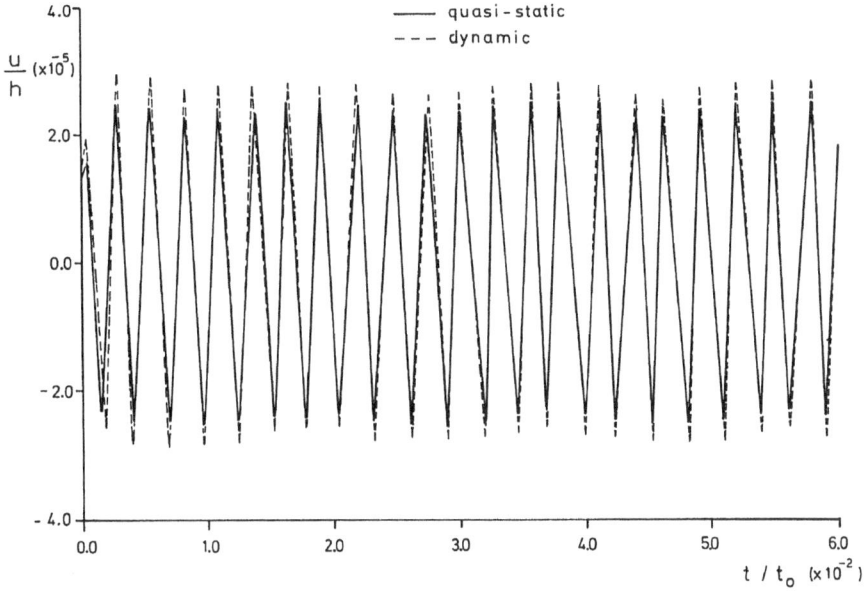

Fig. 9. Solid displacement history under quasi-static and dynamic conditions for the rock column subjected to a harmonic load

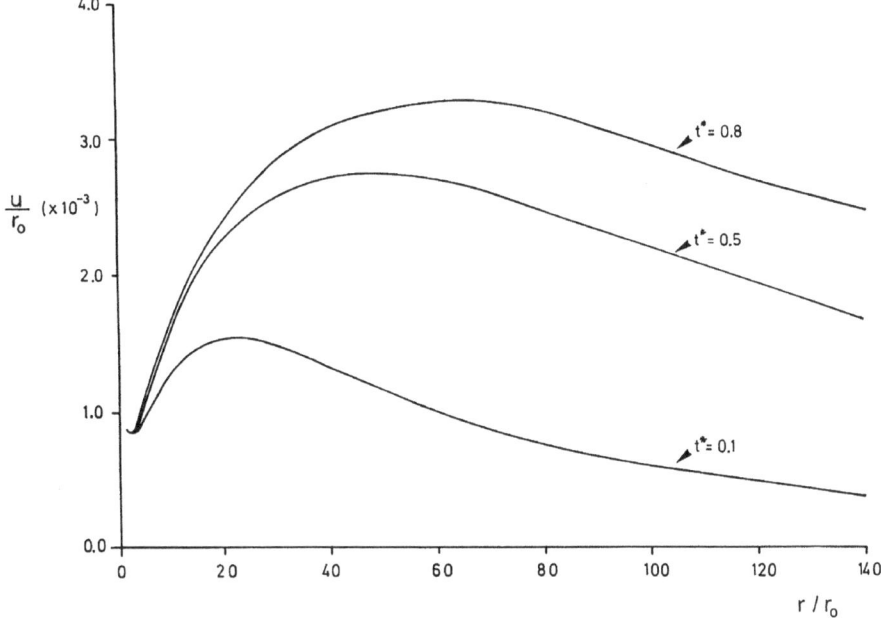

Fig. 10. Solid displacement versus distance for various times $t^* = t/t_0$ in rock borehole problem under suddenly applied pressure

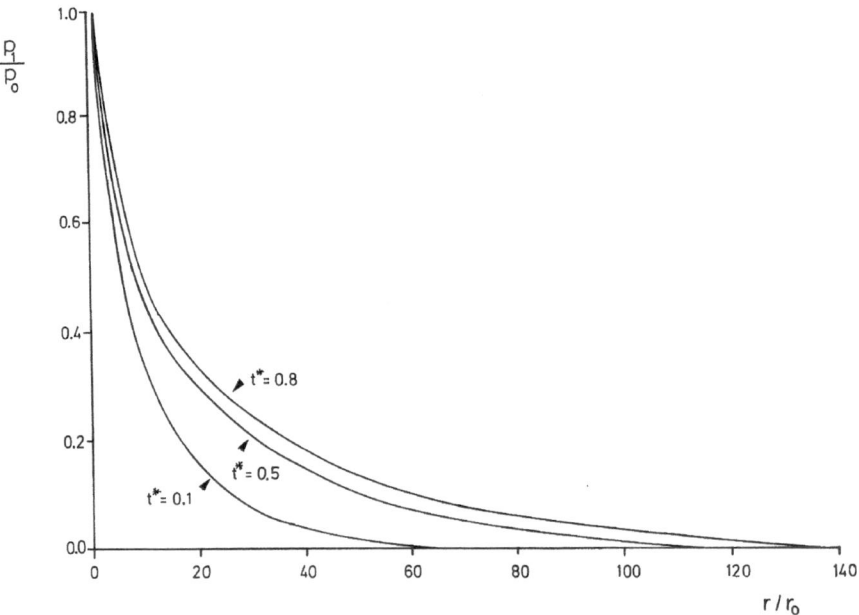

Fig. 11. Fluid fissure pressure versus distance for various times $t^* = t/t_0$ in rock borehole problem under suddenly applied pressure

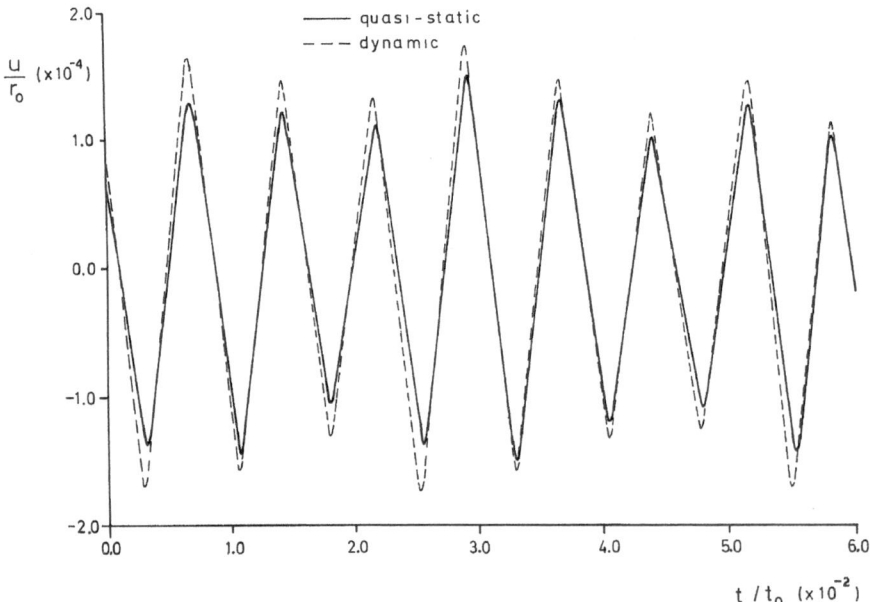

Fig. 12. Solid displacement history under quasi-static and dynamic conditions for the rock borehole subjected to a harmonic pressure

8 Conclusions

On the basis of the preceding discussion the following major conclusions can be drawn:

1) The proposed methodology for the solution of the one-dimensional column and borehole problems in soils and rocks is highly accurate and can easily handle loads of any time variation and soils and rocks exhibiting elastic or viscoelastic material behavior.

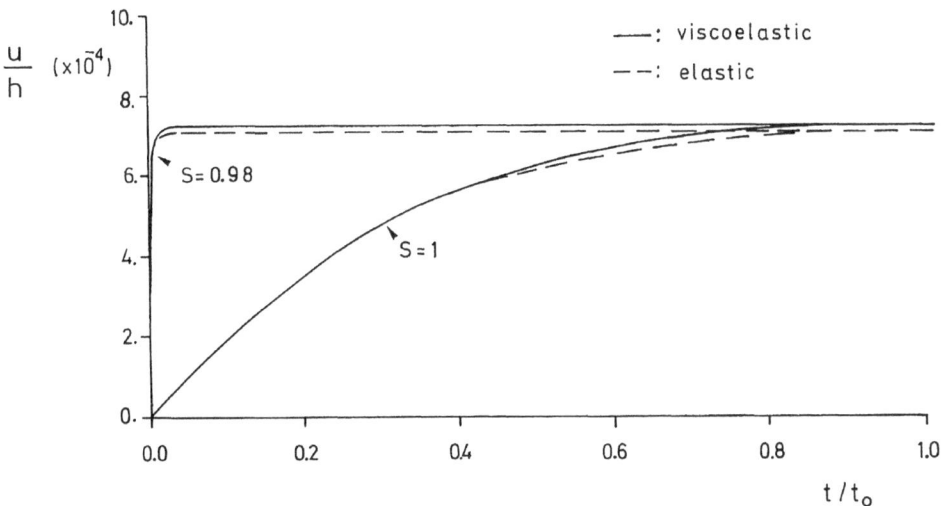

Fig. 13. Solid displacement history for nearly and fully saturated soil column with elastic and viscoelastic skeleton under suddenly applied load

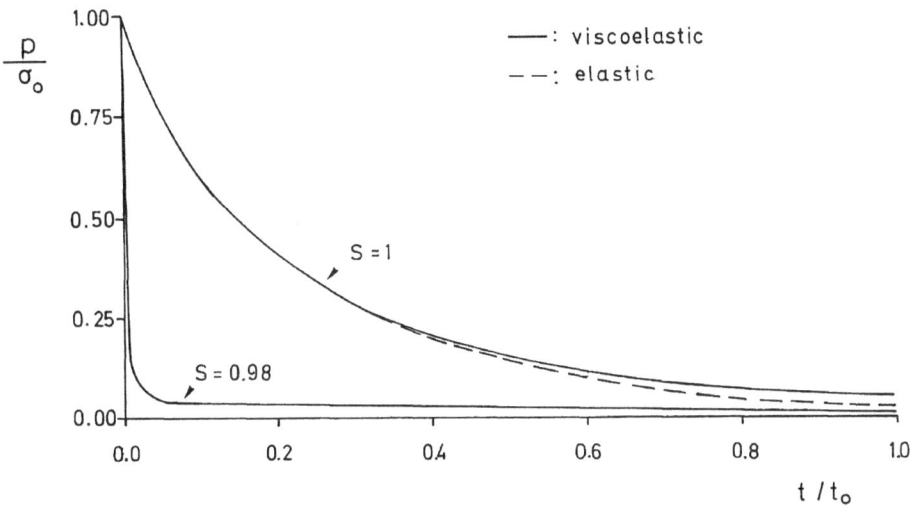

Fig. 14. Fluid pore pressure history for nearly and fully saturated soil column with elastic and viscoelastic skeleton under suddenly applied load

2) The significance of inertial effects depends on the kind of dynamic loading. Thus inertial effects are negligible for suddenly applied loads and significant for harmonic loads, this significance being less pronounced for the case of the borehole problem.

3) The effect of the degree of saturation was found to be significant for the case of the soil model, which takes this effect into account.

4) The effect of porosities and permeabilities in the rock model was found to be rather small and so was found to be the difference between the two degrees and the one degree of porosity rock models.

5) Solution of these one-dimensional column and borehole problems provides valuable insight into the behavior of soils and rocks and a firm basis for checking the accuracy of general numerical methods, such as finite elements or boundary elements.

References

[1] Biot, M. A.: Theory of propagation of elastic waves in a fluid-saturated porous solid. I. low-frequency range. J. Acoust. Soc. Am. **28**, 168–178 (1956).

[2] Bowen, R. M.: Theory of mixtures. In: Continuum physics Vol. III (Eringen, A. C., ed.), pp. 1–127. New York: Academic Press 1976.

[3] Corapcioglu, M. Y.: Wave propagation in porous media — a review. In: Transport processes in porous media (Bear, J., Corapcioglu, M. Y., eds.), pp. 373–469. Dordrecht: Kluwer Academic Publishers 1991.

[4] Vgenopoulou, I., Beskos, D. E.: Dynamic poroelastic soil column and borehole problem analysis. Soil Dynamics and Earthquake Engineering (to appear).

[5] Vgenopoulou, I., Beskos, D. E.: Dynamics of saturated rocks. IV: column and borehole problems. J. Eng. Mech., ASCE (to appear).

[6] Vgenopoulou, I., Beskos, D. E.: Dynamic behavior of saturated poroviscoelastic media. Acta Mechanica (to appear).

[7] Vardoulakis, I., Beskos, D. E.: Dynamic behavior of nearly saturated porous media. Mechanics of Materials **5**, 87–108 (1986).

[8] Aifantis, E. C.: On the problem of diffusion in solids. Acta Mechanica **37**, 265–296 (1980).

[9] Verruijt, A.: Elastic storage of aquifers. In: Flow through porous media (De Wiest, R. J. M., ed.), pp. 331–376. New York: Academic Press 1969.

[10] Verruijt, A.: The theory of consolidation. In: Fundamentals of transport phenomena in porous media (Bear, J., Corapcioglou, M. Y., eds.), pp. 349–368. Dordrecht: Martinus Nijhoff Publishers 1984.

[11] Garg, S. K., Nayfeh, A. H., Good, A. J.: Compressional waves in fluid-saturated elastic porous media. J. Appl. Phys. **45**, 1968–1974 (1974).

[12] Bowen, R. M., Lockett, R. R.: Inertial effects in poroelasticity. J. Appl. Mech. **50**, 334–342 (1983).

[13] Zienkiewicz, O. C., Chang, C. T., Bettess, P.: Drained, undrained, consolidating and dynamic behavior assumptions in soils. Geotechnique **30**, 385–395 (1980).

[14] Simon, B. R., Zienkiewicz, O. C., Paul, D. K.: An analytical solution for the transient response of saturated porous elastic solids. International Journal for Numerical and Analytical Methods in Geomechanics **8**, 381–398 (1984).

[15] Hiremath, M. S., Sandhu, R. S., Morland, L. W., Wolfe, W. E.: Analysis of one-dimensional wave propagation in a fluid-saturated finite soil column. International Journal for Numerical and Analytical Methods in Geomechanics **12**, 121–139 (1988).

[16] Ghaboussi, J., Wilson, E. L.: Variational formulation of dynamics of fluid-saturated porous elastic solids. J. Eng. Mech. Div., ASCE **98**, 947–963 (1972).

[17] Prevost, J. H.: Nonlinear transient phenomena in saturated porous media. Computer Methods in Applied Mechanics and Engineering **30**, 3–18 (1982).

[18] Prevost, J. H.: Wave propagation in fluid-saturated porous media: an efficient finite element procedure. Soil Dynamics and Earthquake Engineering **4**, 183–202 (1985).

[19] Zienkiewicz, O. C., Shiomi, T.: Dynamic behavior of saturated porous media: the generalized Biot formulation and its numerical solution. International Journal for Numerical and Analytical Methods in Geomechanics **8**, 71–96 (1984).

[20] Simon, B. R., Wu, J. S. S., Zienkiewicz, O. C., Paul, D. K.: Evaluation od $u - w$ and $u - \pi$ finite element methods for the dynamic response of saturated porous media using one-dimensional models. International Journal for Numerical and Analytical Methods in Geomechanics **10**, 461–482 (1986).

[21] Simon, B. R., Wu, J. S. S., Zienkiewicz, O. C.: Evaluation of higher order, mixed and hermitean finite element procedures for dynamic analysis of saturated porous media using one-dimensional models. International Journal for Numerical and Analytical Methods in Geomechanics **10**, 483–499 (1986).

[22] Cheng, A. H. D., Badmus, T, Beskos, D. E.: Integral equation for dynamic poroelasticity in frequency domain with boundary element solution. J. Eng. Mech. ASCE **117**, 1136–1157 (1991).

[23] Chang, S. K., Liu, H. L., Johnson, D. L.: Low-frequency tube waves in permeable rocks. Geophysics **53**, 519–527 (1988).

[24] Liu, H. L.: Borehole modes in a cylindrical fluid-saturated permeable medium. J. Acoust. Soc. Am. **84**, 424–431 (1988).

[25] Beskos, D. E.: Dynamics of saturated rocks. I: equations of motion. J. Eng. Mech. ASCE **115**, 982–995 (1989).

[26] Beskos, D. E., Vgenopoulou, I., Providakis, C. P.: Dynamics of saturated rocks. II: body waves. J. Eng. Mech. ASCE **115**, 996−1016 (1989).

[27] Beskos, D. E., Papadakis, C. N., Woo, H. S.: Dynamics of saturated rocks. III: Rayleigh waves. J. Eng. Mech. ASCE **115**, 1017−1034 (1989).

[28] Wilson, R. K., Aifantis, E. C.: On the theory of consolidation with double porosity − I. International Journal of Engineering Science **20**, 1009−1035 (1982).

[29] Beskos, D. E., Aifantis, E. C.: On the theory of consolidation with double porosity − II. International Journal of Engineering Science **24**, 1697−1716 (1986).

[30] Khaled, M. Y., Beskos, D. E., Aifantis, E. C.: On the theory of consolidation with double porosity − III: a finite element formulation. International Journal for Numerical and Analytical Methods in Geomechanics **8**, 101−123 (1984).

[31] Ohnishi, Y., Shiota, T., Kobayashi, A.: Finite element double − porosity model for deformable saturated − unsaturated fractured rock mass. In: Numerical methods in geomechanics Innsbruck 1988 (Swoboda, G., ed.), pp. 765−775. Rotterdam: A. A. Balkema 1988.

[32] Valliappan, S., Khalili-Naghadeh, N.: Flow through fissured porous media with deformable matrix. International Journal for Numerical Methods in Engineering **29**, 1079−1094 (1990).

[33] Biot, M. A.: Theory of deformation of a porous viscoelastic anisotropic solid. J. Appl. Phys. **27**, 459−467 (1956).

[34] Tabaddor, F., Little, R. W.: Constitutive equations for mixtures of Newtonian fluids and viscoelastic solids. Developments in Mechanics **6**, 459−469 (1971).

[35] Podstrigach, Y. S., Pavlina, V. S.: Diffusive processes in a viscoelastic deformed layer. Fizikokhimi-cheskaya Mekhanika Materialov **13**, 76−81 (1977).

[36] Predeleanu, M.: Boundary integral method for porous media. In: Boundary element methods (Brebbia, C. A., ed.), pp. 325−334. Berlin: Springer 1981.

[37] Dziecielak, R.: Propagation of acceleration waves in a fluid-saturated porous viscoelastic solid. Studia Geotechnica et Mechanica **2**, 13−23 (1980).

[38] Taylor, P. A., Aifantis, E. C.: On the theory of diffusion in linear viscoelastic media. Acta Mechanica **44**, 259−298 (1982).

[39] Durbin, F.: Numerical inversion of Laplace transform: an efficient improvement to Dubner and Abate's method. The Computer Journal **17**, 371−376 (1974).

[40] Narayanan, G. V., Beskos, D. E.: Numerical operational methods for time-dependent linear problems. International Journal for Numerical Methods in Engineering **18**, 1829−1854 (1982).

[41] Biot, M. A.: General theory of three-dimensional consolidation. J. Appl. Phys. **12**, 155−164 (1941).

[42] von Rosenberg, D. U.: Methods for the numerical solution of partial differential equations. New York: American Elsevier 1969.

Authors' address: Prof. D. E. Beskos, Department of Civil Engineering, University of Patras, GR-26110 Patras, Greece.

Acta Mechanica (1992) [Suppl] 3: 55–65
© Springer-Verlag 1992

Some advances in boundary integral methods for wave-scattering from cracks

G. Krishnasamy, F. J. Rizzo, and **Y. J. Liu,** Champaign, Illinois, U.S.A.

Summary. This paper deals with some recent and ongoing research involving scattering of time-harmonic acoustic and elastic waves from cracks and cracklike thin scatterers. The character and treatment of the singular integral equations involved in the formulation and solution of such problems are discussed and a number of numerical examples are presented.

1 Introduction

Scattering of elastic waves from arbitrarily – shaped cracks in a linear elastic solid is a topic of continuing interest in solid mechanics, and it is of fundamental importance in the nondestructive evaluation of materials by ultrasonic methods. Formulation of scattering problems in terms of boundary integrals is popular for a number of reasons; these include the ability of such integrals to express the near and far scattered fields with accuracy and stress intensity factors with ease and simplicity. The boundary element method of numerical solution of boundary integral equations, with elements confined to the crack surface, provides an attractive approach to practical problems, especially in three dimensions.

There are two main models for cracks, namely, (i) the mathematical model where the two surfaces of the crack (before loading) occupy the same place, and (ii) a thin-crack model where the two surfaces are distinct but close together. This paper discusses the degeneracy in conventional boundary integral methods for case (i) and the near-degeneracy associated with case (ii), and it suggests certain strategies to surmount these degeneracies.

A scalar counterpart of the crack problem involves scattering of acoustic waves from arbitrarily – shaped, vanishingly – thin rigid screens (cf. [1], [2]). All of the essential ideas and methods involved in the crack problem, that are to be emphasized in this paper, pertain to this scalar problem. Thus, for ease in presentation, in the next section we formulate the scalar version of the crack problem which leads to consideration of a hypersingular integral equation. Then, we discuss a regularization strategy to do computations with this equation followed by some numerical examples in Sect. 4. In Sect. 5 we do consider the vector elastodynamic problem of scattering from a crack and present some data for an elliptical crack. Finally, we consider some of the issues involved with the thin-crack model, suggest a formulation strategy, and present some data for several scalar thin-body scattering problems.

2 Formulation

Consider a time-harmonic acoustic wave incident upon a thin screen as shown in Fig. 1. We are interested in the scattered field which satisfies the Helmholtz equation

$$c^2 \nabla^2 u_s + \omega^2 u_s = 0. \tag{1}$$

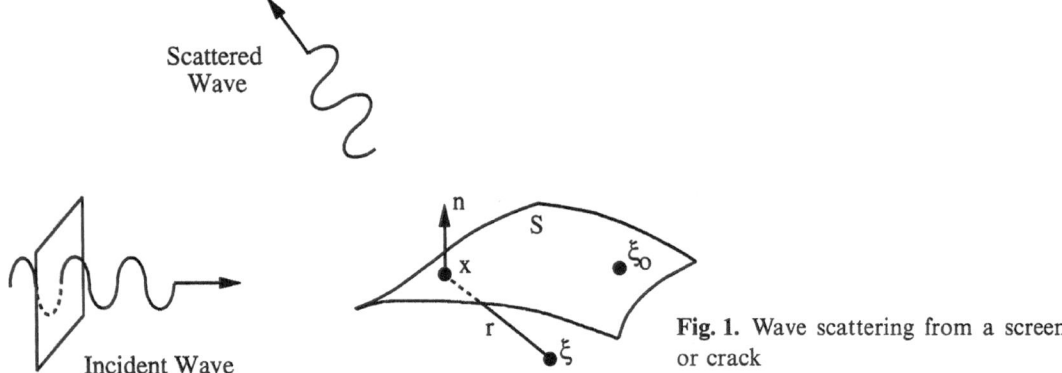

Fig. 1. Wave scattering from a screen or crack

Above, u_s is the scattered acoustic pressure which satisfies the radiation condition at infinity, c is the wave speed, ω is the frequency, and ∇ is the usual 'del' operator in three dimensional space. On the scatterer surface S, $q_s + q^I = q$ with $q_s \equiv \partial u_s/\partial n$ and $q^I \equiv \partial u^I/\partial n$, where u^I is the incident pressure and n is the normal to S (Fig. 1). Incident plus scattered field equals total field (et. seq.).

To find the scattered field, it is expedient to formulate the above problem in terms of integral equations defined on the scatterer surface. In Fig. 2 is shown a thin scatterer S_T with two surfaces S^+ and S^- identified between which there is a small but (as yet) nonzero volume. A familiar application of Green's theorem (cf. [3], [4], [5], [6], [7]) yields the identity

$$\alpha u(\xi) = \oint_{S_T} \left[G(x, \xi)\, q(x) - \frac{\partial G(x, \xi)}{\partial n}\, u(x) \right] dS() + u^I(\xi) \tag{2}$$

where u and q are the total fields; ξ is a point on or off of S_T, x is the location of element of integration dS, G is the free-space Green's function $e^{ikr}/(4\pi r)$, where $k = \omega/c$ and $r = |\xi - x|$, and $\alpha = 1/2$ if ξ is on S_T (at a point ξ_0 with a well defined tangent plane) or $\alpha = 1$ if ξ is off of S but in the acoustic medium (dependence of all quantities on the frequency ω is understood). Now if the scatterer is rigid for example, such that $q = 0$, Eq. (2) with $\alpha = 1/2$ is a boundary integral equation in the unknown u on S_T. Once this equation is solved[1], formula (2) (with $\alpha = 1$), becomes the generator of u at any other ξ, as desired, to complete the solution to the scattering problem.

Next, suppose that the surfaces S^+ and S^- approach each other such that the volume between the surfaces goes to zero (forming a crack in the vector case). Then, there is difficulty with the solution process based on (2) as described above. Specifically, before the limit as S^+ goes to S^-,

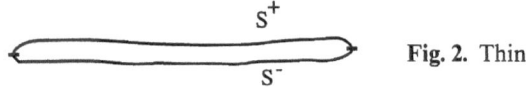

Fig. 2. Thin cracklike scatterer

---

[1] Unique solution to (2) at certain discrete frequencies is impossible, see e.g. [8], [9], [10]. However, modifications of (2) which guarantee a unique solution at all frequencies are available. Nevertheless, for present purposes it is sufficient to note that this uniqueness difficulty disappears as the volume enclosed by S_T goes to zero.

consider the equation

$$\frac{1}{2} u^+({\xi_0}^+) + \int_{S^-} \frac{\partial G(x^+, {\xi_0}^+)}{\partial n^+(x^+)} u^+(x^+) \, dS(x^+) + \int_{S^-} \frac{\partial G(x^-, {\xi_0}^+)}{\partial n^-(x^-)} u^-(x^-) \, dS(x^-)$$

$$= \int_{S^-} G(x^+, {\xi_0}^+) \, q^+(x^+) \, dS(x^+) + \int_{S^-} G(x^-, {\xi_0}^+) \, q^-(x^-) \, dS(x^-) + u^I({\xi_0}^+) \tag{3}$$

and another one just like it except with ${\xi_0}^+$ replaced ${\xi_0}^-$. Each equation represents an application of (2) (with $\alpha = 1/2$) with collocation points $\xi_0 = {\xi_0}^+$ and ${\xi_0}^-$, respectively. Because in the limit $G(x, \xi)$ has identical values for points on S^+ and S^-, and because in the limit $\partial G(x^+, \xi)/\partial n(x^+) = -\partial G(x^-, \xi_0)/\partial n(x^-)$ as well, both Eqs. (3) and its counterpart with $\xi_0 = {\xi_0}^-$ are identical, and each in the limit have the form

$$\frac{1}{2} \Sigma u(\xi_0) + \int_S \frac{G(x, \xi_0)}{\partial n(x)} \Delta u(x) \, dS(x) = \int_S G(x, \xi_0) \, \Sigma q(x) \, dS(x) + u^I(\xi_0) \tag{4}$$

in which $\Delta u = u^+ - u^-$ is the jump in pressure across S^+ and S^-, Σq is the sum $q^+ + q^-$, and S is **either** surface S^+ or S^-. It is apparent now that if, for a rigid scatter, $q^+ = q^- = 0$, and therefore $\Sigma q = 0$, both Δu and Σu are unknown in (4) such that (4) alone is insufficient to obtain either, and the previously described solution process breaks down.

What is usually done to overcome this degeneracy is to take the normal derivative of identity (2), carefully obtain the form of this new identity corresponding to ${\xi_0}^+$ or ${\xi_0}^-$ as with (3), to obtain in the limit the counterpart of (4), i.e.,

$$\frac{1}{2} \Delta q(\xi_0) + \int_S \frac{\partial^2 G(x, \xi_0)}{\partial n(\xi_0) \, \partial n(x)} \Delta u(x) \, dS(x) = \int_S \frac{G(x, \xi_0)}{\partial n(\xi_0)} \Sigma q(x) \, dS(x) + q^I(\xi_0). \tag{5}$$

Now setting Δq and $\Sigma q = 0$ in (5), by virtue of the (rigid) boundary condition, solve (5) for Δu such that Σu across S^+ and S^- may be obtained easily from (4). To obtain u off of the scatterer, ne.g. in the far field, the proper version of (2) (with $\alpha = 1$) to be used is

$$u(\xi) = - \int_S \frac{\partial G(x, \xi)}{\partial n} \Delta u(x) \, dS(x) + u^I(\xi). \tag{6}$$

The use of (5) with (6) and (4) works well for the infinitesimally − thin screen scatterer. References [11], [12], [13], [14], [15] and [16] are merely a sample of the recent work based on this solution strategy. An interesting feature of this strategy is the presence of the term with the 'double dash' through the integral sign in (5). The kernel of this integral involves the second derivative of the Green's function and thus is of the order $1/r^3$. By contrast, the kernel involving the first derivative of the Green's function is of order $1/r^2$, whereas the Green's function itself is of order $1/r$. These kernels are termed hypersingular, strongly or Cauchy singular, and weakly singular, respectively; and the double dash, single dash, and absence of any dash, respectively, are intended to signify the special meanings to be attached to hypersingular and strongly singular integrals (cf. [17−26]). Indeed, quite a body of literature has arisen recently in connection with hypersingular integrals which appear, as described, in the crack problem as well as other contexts (e.g. [27−29], [10]). The strongly singular and weakly singular integrals, however, are comparatively more familiar in boundary integrals analysis.

3 Regularization

There exists a variety of options for regularization (e.g. [30−33]), i.e. lowering the singularity of the integrand of hypersingular integrals before computation is attempted, or for computing them more directly (see the survey in [33]). Here we use and briefly describe only the regularization process involving a two-term Taylor series expansion, as treated more fully in [15].

If Δu is expanded in a Taylor series about ξ_0 and the first two terms are subtracted from Δu in the hypersingular integrand in (5) and added back, it is possible, with the aid of Stokes' theorem, to rewrite the hypersingular integral in (5) in the form

$$
\begin{aligned}
\frac{\partial u^s(\xi_0)}{\partial n(\xi_0)} = -\frac{\partial u^i(\xi_0)}{\partial n(\xi_0)} = {}&-\int_S \left[\frac{\partial^2 G(x, \xi_0)}{\partial n(\xi_0)\, \partial n(x)} - \frac{\partial^2 G^s(x, \xi_0)}{\partial n(\xi_0)\, \partial n(x)} \right] \Delta u(x)\, dS \\[2mm]
&- \int_S \frac{\partial^2 G^s(x, \xi_0)}{\partial n(\xi_0)\, \partial n(x)} \left[\Delta u(x) - \Delta u(\xi_0) - \Delta u_{,p}(\xi_0)\, (x_p - \xi_{0p}) \right] dS \\[2mm]
&- \Delta u_{,k}(\xi_0)\, n_r(\xi_0) \int_S \frac{\partial G^s(x, \xi_0)}{\partial \xi_r}\, n_k(x)\, dS \\[2mm]
&+ \Delta u(\xi_0)\, n_r(\xi_0)\, \varepsilon_{qkr} \oint_C \frac{\partial G^s(x, \xi_0)}{\partial x_k}\, dx_q \\[2mm]
&+ \Delta u_{,p}(\xi_0)\, n_r(\xi_0)\, \varepsilon_{qkr} \oint_C \frac{\partial G^s(x, \xi_0)}{\partial x_k}\, (x_p - \xi_{0p})\, dx_q \\[2mm]
&+ \Delta u_{,p}(\xi_0)\, n_r(\xi_0)\, \varepsilon_{qrp} \oint_C G^s(x, \xi_0)\, dx_q
\end{aligned}
\tag{7}
$$

wherein C is the (line) boundary of S and G^s is the static Green's function. The subtraction of G^s from G as shown is done for convenience such that the kernel in most terms is independent of frequency. No integral in (7) is more than weakly singular. Conventional boundary element methods of solving the integral equations, in use for years, may be used with proper care given to the smoothness demanded by (7) [34].

4 Numerical examples

For illustration, consider a penny-shaped rigid scatterer. The scatterer surface S^+ or S^- is discretized into 25 elements using three rings and 8 radial lines with one (circular) element in the middle. These elements used to describe the scatterer are the standard conforming quadratic elements [33], but the jump in potential, Δu, on the scatterer surface is approximated by nonconforming elements, where the collocation points are away from the element edges and have sufficient smoothness for (7) to exist. The square-root behavior of the solution along the crack edges is built into the elements at the crack edge. The Δu for inclined waves at $ka = 3$ and $ka = 4$ at $30°$ and $45°$ with the normal, respectively, is as in Figs. 3 and 4. These data are verified by

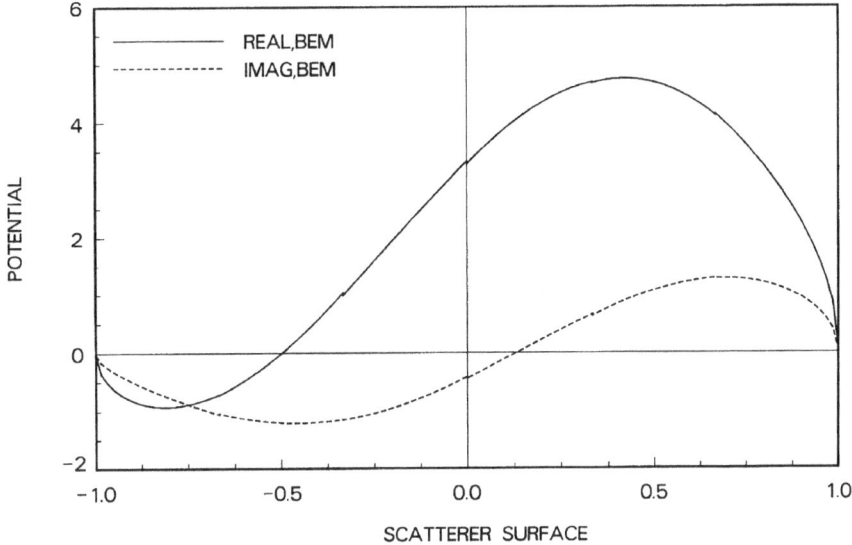

Fig. 3. Scattering due to plane wave inclined at $30°$ to the normal and $ka = 3.0$

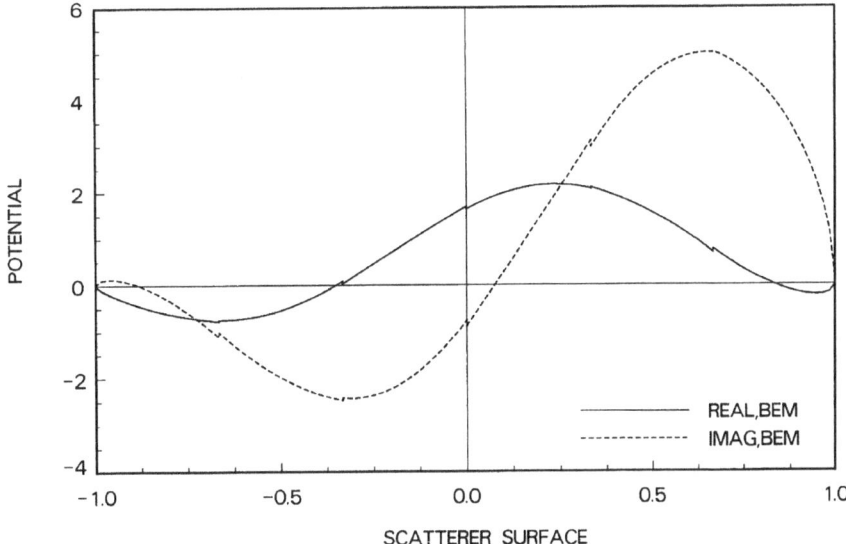

Fig. 4. Scattering due to plane wave inclined at $45°$ to the normal and $ka = 4.0$

solving the problems again using the essentially independent thin-body formulation as described in Sect. 6. Also, for normally incident waves, the data (see [15]) compare well with the analytical solution.

5 Vector crack problem − examples

The scattering of elastic waves from an arbitrarily shaped crack can be expressed as an integral equation,

$$-t_p{}^I(\boldsymbol{\xi}_0) = C_{pqmr}n_q(\boldsymbol{\xi}_0) \left\{ -C_{ijkl} \int\limits_S \left[\frac{\partial^2 G_{km}(\boldsymbol{x}, \boldsymbol{\xi})}{\partial \xi_r \, \partial x_l} - \frac{\partial^2 G^s_{km}(\boldsymbol{x}, \boldsymbol{\xi})}{\partial \xi_r \, \partial x_l} \right] n_j(\boldsymbol{x}) \, \Delta u_i(\boldsymbol{x}) \, dS \right.$$

$$- C_{ijkl} \int\limits_S \frac{\partial^2 G^s_{km}(\boldsymbol{x}, \boldsymbol{\xi})}{\partial \xi_r \, \partial x_l} \, n_j(\boldsymbol{x}) \, [\Delta u_i(\boldsymbol{x}) - \Delta u_i(\boldsymbol{\xi}_0) - \Delta u_{i,p}(\boldsymbol{\xi}_0) \, (x_p - \xi_{0p})] \, dS$$

$$+ C_{ijkl} \int\limits_S \frac{\partial G^s_{im}}{\partial x_r} \, \Delta u_{k,l}(\boldsymbol{\xi}_0) \, n_j(\boldsymbol{x}) \, dS + \Delta u_i(\boldsymbol{\xi}_0) \, \varepsilon_{jrq} C_{ijkl} \oint\limits_C \frac{\partial G^s_{km}}{\partial x_l} \, dx_q$$

$$+ \Delta u_{i,j}(\boldsymbol{\xi}_0) \, \varepsilon_{rlq} C_{ijkl} \oint\limits_C G^s_{km} \, dx_q + \Delta u_{i,p}(\boldsymbol{\xi}_0) \, \varepsilon_{jrq} C_{ijkl} \oint\limits_C \frac{\partial G^s_{km}}{\partial x_l} \, (x_p - \xi_{0p}) \, dx_q$$

$$+ \left[\frac{\Omega(\boldsymbol{\xi}_0)}{4\pi} \right] \Delta u_{m,r}(\boldsymbol{\xi}_0)$$

$$- \frac{1}{8\pi} \, \Delta u_{j,r}(\boldsymbol{\xi}_0) \oint\limits_C \varepsilon_{mji} r_{,pp} \, dx_i - \frac{1}{8\pi(1-v)} \, \Delta u_{j,r}(\boldsymbol{\xi}_0) \oint\limits_C \varepsilon_{jpi} r_{,pm} \, dx_i \left. \right\} \qquad (8)$$

where Δu_i is the crack opening displacement across the crack surface, t^I is the traction due to the incident field, G^D_{km} is the free-space, time-harmonic, elastodynamic Green's function and G^s_{km} is

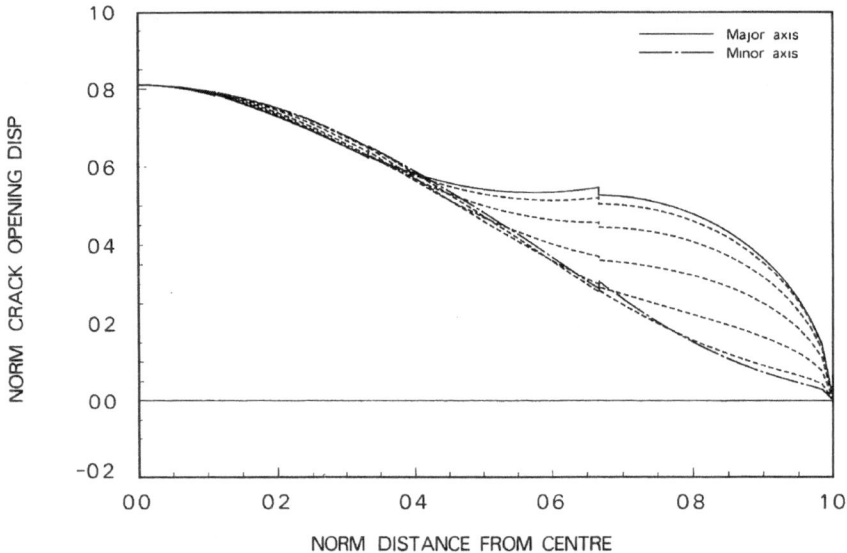

Fig. 5. Normalized crack opening displacement for elliptical crack, $k_t a = 4.0$, *No.* of elements $= 25$

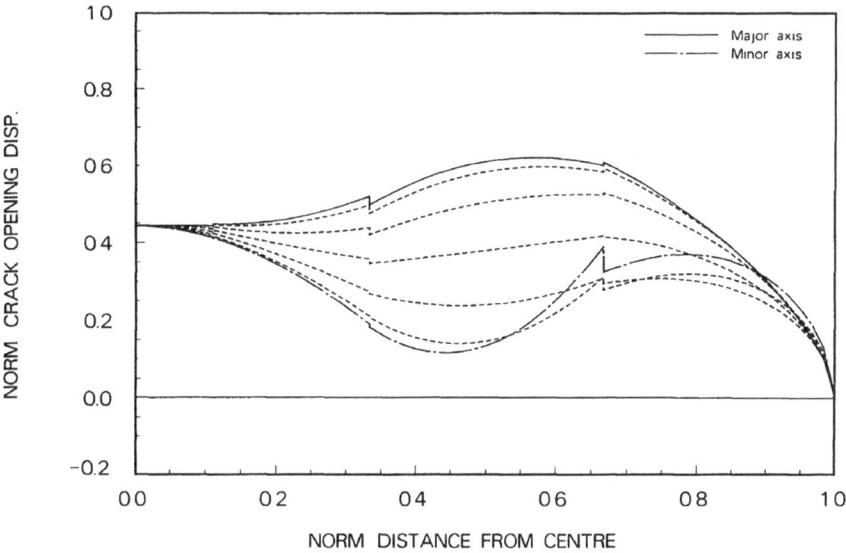

Fig. 6. Normalized crack opening displacement for elliptical crack, $k_t a = 5.0$, *No.* of elements $= 25$

the static equivalent of G_{km}^D otherwise referred to as the Kelvin solution (cf. [15]). Again C is the line which encloses the crack. For flat circular or elliptical cracks, the above equation can be solved for Δu_i. When a plane wave at $k_t a = 4.0$ and $k_t a = 5.0$ strikes an elliptical crack of aspect ratio 2 at normal incidence, Δu_n, the crack opening displacement normal to the crack surface is presented in Figs. 5 and 6. In these figures the crack opening displacement is plotted along the major and minor axis and radial lines at 15 degree intervals between the major and minor axis. These data compare well with those presented in [14].

6 Thin scatterer (cracklike) model

We return now to the scalar problem of scattering of acoustic waves, but here we examine a different scatterer, namely, a thin rigid scatterer of small but finite thickness. Our intention is to come to grips with the two main difficulties associated with the thinness of such shapes. These are (a) ill-conditioning of the formulation (i.e. (3) and its counterpart with ξ_0^+ replaced by ξ_0^-) which was shown to degenerate in the limit as the two scatterer surfaces come together, and (b) difficulty in obtaining accurate values of the integrals over boundary elements which are very close together across the small thickness. Such problems are important in a variety of cases where the arbitrarily − thin or true-crack model is inappropriate. They are useful also in testing the limits of applicability of the simpler true-crack model.

Therefore, to begin, consider the normal derivative of identity (2) again but now applied to the thin shape of Fig. 2 with specific reference to (collocation) points ξ_0^- as written below

$$\frac{1}{2} q^-(\xi_0^-) + \int_{s^+} \frac{\partial^2 G(x^+, \xi_0^-)}{\partial n^+(x^+)\, \partial n^-(\xi_0^-)}\, u^+(x^+)\, dS(x^+) + \int_{s^-} \frac{\partial^2 G(x^-, \xi_0^-)}{\partial n^-(x^-)\, \partial n^-(\xi_0^-)}\, u^-(x^-)\, dS(x^-)$$

$$= \int_{s^+} \frac{\partial G(x^+, \xi_0^-)}{\partial n^-(\xi_0^-)}\, q^+(x^+)\, dS(x^+) + \int_{s^-} \frac{\partial G(x^-, \xi_0^-)}{\partial n^-(\xi_0^-)}\, q^-(x^-)\, dS(x^-) + q^I(\xi_0^-). \tag{9}$$

The combination of (9) with formula (3) has very different properties compared with the combination of (3) with its counterpart referring to point $\xi_0{}^-$. Indeed, and it is a key observation for thin-body problems, that a formulation involving a conventional boundary integral equation (e.g. (3)) collocated on one surface of a thin shape and with a hypersingular boundary integral equation (e.g. (9)) collocated on the other surface is nondegenerate in the limit and therefore well-conditioned for thin shapes. This observation takes care of difficulty (a) above.

Difficulty (b) remains for thin shapes regardless of the formulas (conventional or hypersingular) used. However, it too can be surmounted, and the specific means for doing so is addressed in another paper [35] which deals in more detail with the formulation of the thin-body problem as well as more detail with numerical issues (see also [36]). It suffices here to mention that the key ingredient in developing our integration scheme, accurate even for closely-parallel boundary elements, involves reduction of all integrals to weakly singular form. This is accomplished through the two-term Taylor series expansion for the density functions in the highly singular integrands, as mentioned in Sect. 3.

7 Thin scatterer — examples

Consider the problem of scattering of acoustic waves by a thin rigid screen of thickness $2h$ as shown in Fig. 7. Waves of various frequencies and angles of incidence ϕ impinge upon the scatterer and we are interested in the scattered field for various thicknesses h. Our purpose is to examine the conditioning of our equations and the accuracy of our solution as h gets small. In the process we can compare differences between the thin body and the arbitrarily — thin model of the scatterer. To do the computations, we discretize both surfaces S^+ and S^- of the circular scatterer with elements as used for only one surface in Sect. 4.

In Fig. 8 is shown the magnitude of the scattered field for $ka = 1$, at a distance of 5 radii, as a function of θ, for a normal-incident wave, for various values of h compared with the 'arbitrarily' thin $h = 0$ model. The formulation is well conditioned at the values of h shown as well as smaller values down to $h = 10^{-7}$. Note for values of h less than about 0.1, there is little difference with the $h = 0$ model. We should mention that the $h = 0$ model has built in the square root singularity at the edges whereas the finite thickness model does not.

Therefore, for $h = 0.1$ we examined the backscatter and specular scatter as a function of incident-wave angle at various values of ka and compared with the $h = 0$ model at various distances from the crack. Specific data are not shown but in essence we found that differences with the $h = 0$ model decrease with distance and with higher frequency.

We close this section with the observation that if the thin-body shape is an inclusion, i.e. a region of fluid of (perhaps) different properties rather than a rigid inclusion, it is possible to solve this problem using a formulation with the same good properties as found above. Indeed, as a check, we modelled a thin inclusion with the same properties as the surrounding field, and obtained the homogeneous (i.e. no scatterer) solution, as expected.

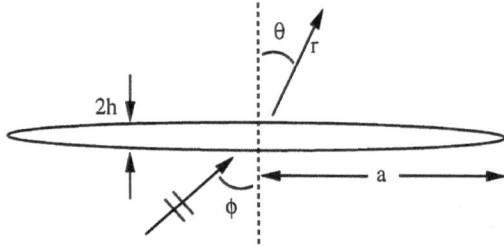

Fig. 7. Edge view of thin circular rigid scatterer

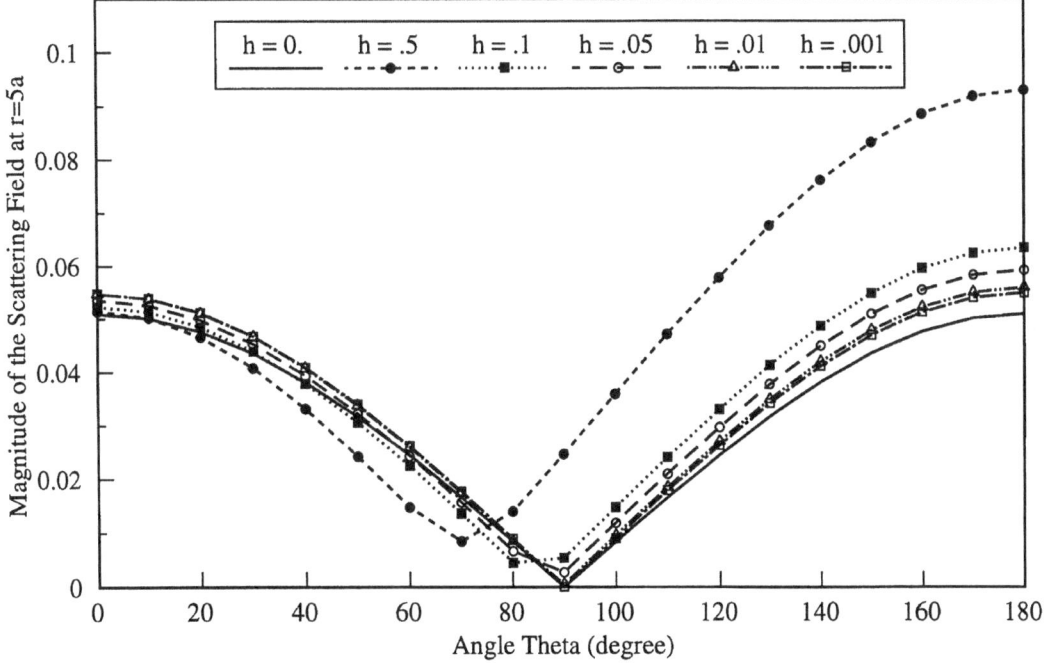

Fig. 8. Scattering for a normal incident wave, N_0 of elements $= 10$, N_0 of nodes $= 80$, $ka = 1.0$

8 Discussion

In this paper we have attempted to digest some of our recent and ongoing research with integral methods for scattering from cracks in solids and cracklike shapes in acoustic fluids. Specifically, we indicated the degeneracy in the conventional formulation for cracks or arbitrarily – thin scatterers and presented an appropriate remedy involving hypersingular integral equations. Data for some example problems for scalar and vector waves are presented following a particular strategy for dealing with (regularizing) the hypersingular integrals. Then the problems associated with scattering from thin bodies or cracklike scatterers with associated near degeneracy was addressed. Here we suggested a formulation based on collocation with conventional equations on one surface of the thin body and collocation with hypersingular equations on the other surface. Such a formulation has excellent properties and an illustrative example involving acoustic scattering from a thin penny-shaped rigid scatterer was presented for verification.

In closing we should point out some recent work [37 – 39] for scattering from multiple objects which involves a series approach. In this work the influence of the interactive scattering between scatterers is associated with the number of terms in a series. Thus, truncating the series neglects higher order interactive scattering. The method is shown [39] to be effective and efficient for a number of shapes, including those which are cracklike, even though the formulation involves conventional integral equations (as supposed to hypersingular) only.

All of these treatments of scattering from cracks, cracklike shapes and other shapes, in fluids or solids may be regarded as ways of solving the so called forward problem, i.e. the problem where the scatterers are known and the scattered field is unknown. The more difficult problem, and the one more technologically significant in nondestructive evaluation and target identification and characterization, is the inverse problem, wherein the object doing the scattering is unknown. It is well understood, however, that almost all strategies for the inverse problem require accurate data and efficient schemes for solving the forward problem. Indeed, boundary integral methods are already used in that capacity e.g. [40], [41]. Hopefully, the work developed above may ultimately find its way as a valued ingredient in such inverse problem strategies.

Acknowledgement

Portions of this work were supported by the National Science Foundation under grant No. NSF MSS-8918005, by the College of Engineering, and by the Theoretical and Applied Mechanics Department at the University of Illinois.

References

[1] Martin, P. A., Rizzo, F. J.: On boundary integral equations for crack problems. Proc. Roy. Soc. A 421, 341—355 (1989).

[2] Rudolphi, T. J., Krishnasamy, G., Schmerr, L. W., Rizzo, F. J.: On the use of strongly singular integral equations for crack problems. Boundary elements 10 (C. A. Brebbia, ed.), Southampton, England: Springer-Verlag and Computational Mechanics Publications. Sept. 1988.

[3] Jaswon, M. A.: Integral equations methods in potential theory. Proc. Royal. Soc. A 275, 23—32 (1963).

[4] Banerjee, P. K., Butterfield, P. K.: Boundary element method in engineering. McGraw-Hill 1981.

[5] Mukherjee, S.: Boundary element method in creep and fracture. New York: Applied Sciences Publishers 1982.

[6] Cruse, T. A.: Boundary element analysis in computational fracture mechanics. Boston: Kluwer Academic Publishers 1988.

[7] Hartmann, F.: Introduction to boundary element theory and applications. New York: Springer 1989.

[8] Schenck, H. A.: Improved integral formulation for acoustic radiation problems. J. Acoust. Soc. Am. 44, 41—58 (1968).

[9] Kleinman, R. E., Roach, G. F.: Boundary integral equations for the three-dimensional Helmholtz equation. SIAM Rev. 16, 214—236 (1974).

[10] Liu, Y. J., Rizzo, F. J.: A weakly-singular form of the hypersingular boundary integral equation applied to 3-D acoustic wave problems. Computer methods in applied mechanics and engineering. To appear.

[11] Takakuda, K., Koizumi, T., Shibuya, T.: On integral equation methods for crack problems. Bull. J.S.M.E. 28, 236 (1985).

[12] Nishimura, N., Kobayashi, S.: An improved boundary integral equation method for crack problems. Proc. IUTAM Symposium, San Antonio, Texas, USA, April 13—16, 1987 (T. A. Cruse, ed.), Adv. Boundary Element Method. Berlin—Heidelberg—New York—Tokyo: Springer 1987.

[13] Lin, W., Keer, L. M.: Scattering by a planar three-dimensional crack. J. Acoust. Soc. Am. 82 (4), 1442—1448 (1987).

[14] Budreck, D. E., Achenbach, J. D.: Scattering from three-dimensional planar cracks by the boundary integral equation method. J. Appl. Mech. 55, 405—411 (1988).

[15] Krishnasamy, G., Schmerr, L. W., Rudolphi, T. J., Rizzo, F. J.: Hypersingular boundary integral equations: Some applications in acoustic and elastic wave scattering. J. Appl. Mech. 57, 404—411 (1990).

[16] Gray, L. J., Martha, L. F., Ingraffea, A. R.: Hypersingular integrals in boundary element fracture analysis. Int. J. of Num. Meth. in Engg. 29, 1135—1158 (1990).

[17] Muskhelishvili, N. I.: Singular integral equations. Nordhoff: The Netherlands 1953.

[18] Hadamard, J.: Lectures on cauchy's problem in linear partial differential equations. New Haven, Conn.: Yale Univ. Press 1923.

[19] Kutt, H. R.: The numerical evaluation of principal value integrals by finite-part integration. Numer. Math. 24, 20—210 (1975).

[20] Brando, M. P.: Improper integrals in theoretical aerodynamics: The problem revisited. AIAA J. 25 (9), 1258—1260 (1986).

[21] Kaya, A. C., Erdogan, F.: On the solution of integral equations with strongly singular kernels. Q. Appl. Math. 57, 404—414 (1987).

[22] Ioakimidis, N. I.: A new singular integral equation for the classical crack problem in plane and antiplane elasticity. Int. J. Fracture **21**, 115−122 (1983).

[23] Guiggiani, M., Gigante, A.: A general algorithm for multidimensional cauchy principal value integrals in the boundary element method. J. Appl. Mech. **57**, 906−915 (1990).

[24] Ervin, V. J., Kieser, R., Wendland, W. L.: Numerical approximation of the solution for a model 2-D hypersingular integral equation. Computational engineering with boundary elements, Vol. 1 (S. Gilli, C. A. Brebbia, A. H. D. Cheng, eds.), Computational Mechanics Publications, Southampton 1990.

[25] Guiggiani, M., Krishnasamy, G., Rudolphi, T. J., Rizzo, F. J.: A general algorithm for the numerical solution of hypersingular boundary integral equations. J. Appl. Mech., to appear.

[26] Schwab, C., Wendland, W. L.: Numerical integration of singular and hypersingular integrals in boundary element methods. Numerische Mathematik, in review.

[27] Filippi, P. J. T.: Layer potentials and acoustic diffraction. J. Sound Vibr. **54 (4)**, 473−500 (1977).

[28] Nedelec, J. C.: Integral equations with non integrable kernels. Integral Eq. Operator Th. **5**, 562−572 (1982).

[29] Chien, C. C., Rajiyah, H., Atluri, S. N.: An effective method for solving the hypersingular integral equation in 3-D acoustics. J. Acoust. Soc. Am. **88 (2)**, 918−936 (1990).

[30] Sladek, V., Sladek, J., Balas, J.: Boundary integral formulation of crack problems. Z. Angew. Math. Mech. **66 (2)**, 83−94 (1986).

[31] Bonnet, M.: Regular boundary integral equations for three-dimensional finite or infinite bodies with and without curved cracks in elastodynamics. Boundary element techniques: Applications in engineering, Brebbia, C. A. and Zamani, N. (eds.). Computational Mechanics Publications, Southampton 1989.

[32] Rudolphi, T. J.: The use of simple solutions in the regularization of hypersingular boundary integral equations. Math. Comput. Modelling **15 (3−5)**, 269−278 (1991).

[33] Krishnasamy, G., Rizzo, F. J., Rudolphi, T. J.: Hypersingular boundary integral equations: Their occurrence, interpretation, regularization and computation. To appear in developments in boundary element methods, Vol. 7, Advanced Dynamic Analysis, Elsevier Applied Science Pub.

[34] Krishnaamy, G., Rizzo, F. J., Rudolphi, T. J.: Continuity requirements for density functions in the boundary integral equation method. Computational Mechanics, 9, 267−284 (1992).

[35] Krishnasamy, G., Rizzo, F. J., Liu, Y. J.: Boundary integral equation formulation for thin body problems. In preparation.

[36] Krishnasamy, G., Rizzo, F. J., Liu, Y.: Scattering of acoustic and elastic wave scattering by crack like objects: The role of hypersingular integrals. Review of progress in quantitative nondestructive evaluation (Thompson, D. O., Chimenti, D. E., eds.), Vol. 18. Maine: Plenum Press, Aug. 1991.

[37] Schuster, G. T.: A hybrid BIE + Born series modeling scheme: Generalized Born series. J. Acoust. Soc. Am. **77**, 865−886 (1985).

[38] Kitahara, M., Nakagawa, K.: Born series approach applied to three dimensional elastodynamic inclusion analysis by BIE methods. Proc. Fourth Japan Nat. Symp. on Bound. Elem. Meth., Japan. Soc. Comp. Meth. Engr., Tokyo (1987).

[39] Schafbuch, P. J., Rizzo, F. J., Thompson, R. B.: Elastic wave scattering by multiple inclusions. Proc. ASME Symp. on Enhancing Analysis Techniques for Composite Materials, Winter Annual Meeting, Atlanta, Dec. 1991.

[40] Nishimura, N., Kobayashi, S.: Further applications of regularised integral equations in crack problems. Symp. of the Int. Ass. for Boundary Element Method, Rome, Italy, Oct. 1990.

[41] Kitahara, M., Nakagawa, K.: Phase shift determination of scattered far-fields and its application to an inverse problem. Review of progress in quantitative nondestructive evaluation (Thompson, D. O., Chimenti, D. E., eds.), Vol. 18. Maine: Plenum Press, Aug. 1991.

Authors' addresses: Dr. Guna Krishnasamy, PDA Engineering, 2975 Redhilll Ave., Costa Mesa, CA 92 626, Yijun Liu, Dept. of Theoretical and Applied Mechanics, University of Illinois, Urbana, IL 61 801, Prof. Frank J. Rizzo, Department of Aerospace Enginering and Engineering Mechanics, Iowa State University, Ames, Iowa 50 011, U.S.A.

Acta Mechanica (1992) [Suppl] 3: 67−82

On the propagation of small perturbations in viscous compressible fluid

C. Ferrari, Torino, Italy

Summary. Object of this paper is to prove that the paradox of the instantaneous propagation of small perturbations in the flow of a compressible viscous fluid is removed if one takes the relation between the stress tensor and the deformation rate tensor not the one given by the classical theory but that obtained considering the dependence on the time of the distribution function ϕ of the molecular velocities. The result is derived either determining the momentum transfer due to the thermal molecular motion in analogous way as that of C. Cattaneo in the problem of the heat transfer, or assuming a form of Φ on the ground of invariant considerations according to the method used by M. M. Brillouin. The general motion equations corresponding to the constitutive equation so obtained are written and applied to the case of slow motion and initial value problem (Cauchy problem); the motion equations in this case can be reduced to a system of three partial differential equations of the first order, that results to be totally hyperbolic. The characteristic lines are deduced as well as the variation laws of all the physical quantities along them, and it is shown how the solution can be obtained by an iterative method.

1 Introduction

It is usually assumed that the equations of the slow motion of a viscous compressible fluid are given by the Stokes-Navier equations, and actually their application to a very great variety of problems leads to results that are in good agreement with the experimental ones; on the other hand it is also known that said equations lead to the impossibility of wave propagation (P. Duhem [1], T. Levi-Civita [2]), so that a small perturbation produced in a whatever point in any moment is instantly felt in any other point of the field, even if with a different intensity from point to point. C. Possio [3] gave an ingenious representation of the way in which the sound propagation occurs, when one applies said equations studying the flow that, when the waves exist, it is called "flow by plane waves": according to it, the (small) velocity variation which is produced initially in a point and that in a non-viscous fluid is propagated by a plane wave with the sound speed, in a viscous fluid is propagated "practically" by a plane layer that advancing spreads and whose thickness increases proportionally to the square-root of the time, while its median line advances with a velocity equal to the sound speed as in the non-viscous case: "practically" here means that in the strict sense the velocity variation downstream of the layer is not zero, becoming really nil only at infinity, but it has values relatively very small. From this representation it appears clearly enough which are the limits of the approximation to the real phenomenon corresponding to the application of the Stokes-Navier equations to the problem above indicated, and, on the other hand, it does not remove the absurdity, at least from the conceptual point of view, of the instantaneous propagation of any small perturbation. It is known that an analogous paradoxical result is obtained in the heat transfer problem if one accepts, as usually, the Fourier law to write the constitutive equation linking the heat flux density to the thermal gradient. Now, in this case, C. Cattaneo [4] pointed out that such a paradox is removed if one takes in account that the distribution function of the molecular velocities in a second order theory is an explicit function of the time. It seems therefore not devoid of interest to try to see if also for the acoustic phenomenon the propagation velocity turns out to be finite if one takes the relation between the stress and deformation rates tensors not that given by the classical theory, but that derived taking in account the dependence on the time of the distribution function Φ. This is just the object of this paper, and the result is obtained either making a non-banal extension of the method used by Cattaneo (non-banal because now the quantity which is transferred by the thermal molecular motion is a vector quantity-momentum- and not a scalar one-temperature-), or assuming a form of Φ on the ground of invariant considerations according to the procedure used by M. M. Brillouin [5]. A constitutive equation is obtained which has an additional term to that given by the classical theory; the corresponding dynamical equations of the motion are partial differential equations of second order with respect to the time t, instead of being of the first order.

Applying such equations to the case of motion, that for a non-viscous fluid is corresponding to the flow by plane waves, the problem is reduced to a system of three partial differential equations of the first order in two independent variables, which turns out to be totally hyperbolic. To this system belong therefore three families of characteristic curves which allow easily to calculate numerically the solution of the problem, as well as to determine the domain of dependence of any point P on the initial data, and the domain of influence of these data. It is noteworthy the property that, although the additional term in the constitutive equation is very small with respect to the classical one the propagation velocity of the waves, $c_s{}^*$ is very near to that c_s of the non-viscous fluid; if in the flow equations $c_s{}^*$ is taken just equal to c_s these equations can be easily integrated, while in the case in which one takes in account the difference between $c_s{}^*$ and c_s the characteristic equations allow not only to obtain a numerical solution, but also to derive a non numerical solution by means of an iterative method.

2 Derivation of the constitutive equation

Let us consider first of all a gas flow that, with reference to a system of orthogonal axes x_i ($i = 1, 2, 3$) has a velocity u_1 which is a function of x_2 only, and it is directed along the x_1 axis: $u = u_1(x_2, t)\, i_1$.

In this case the procedure followed by C. Cattaneo [4] can been identically applied. Consider the molecules, whose velocity has a value between c and $c + dc$ and a direction between θ and $\theta + d\theta$ being θ the angle between the velocity c and the positive axis x_2: the number per unit volume $dn_{c\theta}$ of such molecules that cross in the time dt the element of area dS in the plane (x_1, x_3) is given by the relation:

$$dn_{c\theta} = \frac{c}{2} \sin \theta \cos \theta \, d\theta \, dt \, dN_c \tag{1}$$

if dN_c is the total number of molecules per unit volume whose velocity lies between c and $c + dc$. Let be l the distance travelled by anyone of molecules between the moment in which its last collision with the other molecules occurred, and the moment (t_0) of the crossing; that collision happened therefore at the value of x_2 given by $x_2 = -l \cos \theta$ (Fig. 1), and if one admits that the

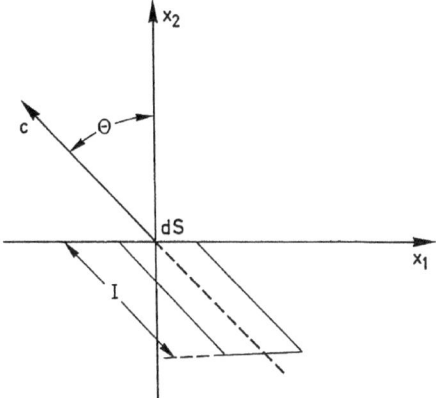

Fig. 1. Transfer of momentum in case I: $u = u_1(x_2, t)\, i_1$

molecules in question have in average a component G_{x_1} of momentum along x_1 equal to the mean value of such quantity in the plane from which they come at the moment $t = t_0 - \dfrac{l}{c}$ this mean value results to be:

$$
\left[\langle G_{x_1}\rangle - l\cos\theta\,\frac{\partial\langle G_{x_1}\rangle}{\partial x_2} - \frac{l}{c}\frac{\partial}{\partial t}\langle G_{x_1}\rangle + \frac{1}{2}\,l^2\cos^2\theta\,\frac{\partial^2\langle G_{x_1}\rangle}{\partial x_2{}^2} \right.
$$
$$
\left. + \frac{l^2\cos\theta}{c}\frac{\partial^2}{\partial t\,\partial x_2}\langle G_{x_1}\rangle + \frac{1}{2}\frac{l^2}{c^2}\frac{\partial^2}{\partial t^2}\langle G_{x_1}\rangle \right]_{x_2 = 0}.
\tag{2}
$$

Summing the contributions, given by all the molecules of the class above considered in an interval of time dt, and dividing by dt, one obtains that the component of momentum along the axis x_1 transferred by them through dS in unit time and per unit area is

$$
\left[\langle G_{x_1}\rangle - \lambda_c\cos\theta\,\frac{\partial}{\partial x_2}\langle G_{x_1}\rangle - \frac{\lambda_c}{c}\frac{\partial}{\partial t}\langle G_{x_1}\rangle + \frac{1}{2}\frac{l_c{}^2}{c^2}\frac{\partial^2}{\partial t^2}\langle G_{x_1}\rangle \right.
$$
$$
\left. + \frac{1}{2}\,l_c{}^2\cos^2\theta\,\frac{\partial^2}{\partial x_2{}^2}\langle G_{x_1}\rangle + \frac{l_c{}^2\cos\theta}{c}\frac{\partial^2}{\partial x_2\,\partial t}\langle G_{x_1}\rangle \right] dn_{c\theta}
\tag{3}
$$

where it has been taken $\sum l = \lambda_c\,dn_{c\theta}$ being the sum extended over all the molecules above indicated, and λ_c the mean-free path of the molecules whose velocity is c, while l_c is the mean-square value of the length l. By integration with respect to θ, being θ variable in the range $0 \div \pi/2$ one obtains the momentum transferred by the molecules having velocities $c \div c + dc$:

$$
\left[c\,\frac{\langle G_{x_1}\rangle}{4} - c\,\frac{\lambda_c}{6}\frac{\partial}{\partial x_2}\langle G_{x_1}\rangle - \frac{\lambda_c}{4}\frac{\partial}{\partial t}\langle G_{x_1}\rangle + c\,\frac{l_c{}^2}{16}\frac{\partial^2}{\partial x_2{}^2}\langle G_{x_1}\rangle \right.
$$
$$
\left. + \frac{l_c{}^2 6}{\partial x_2\,\partial t}\langle G_{x_1}\rangle + \frac{l_c{}^2}{8c}\frac{\partial^2}{\partial t^2}\langle G_{x_1}\rangle \right] dN_c.
\tag{4}
$$

A similar expression is obtained for the molecules crossing dS in the direction of negative x_2 axis; in that expression all terms have the same sign as in (5), except for those in which the differentiation with respect to x_2 is of the first order, and these on the contrary have the opposite sign.

It turns out that the momentum flux Q_x is

$$Q_x = \left[\frac{-c\lambda_c}{3} \frac{\partial}{\partial x_2} \langle G_{x_1} \rangle + \frac{l_c^2}{3} \frac{\partial^2}{\partial x_2 \partial t} \langle G_{x_1} \rangle \right] dN_c. \tag{5}$$

Integrate with respect to c, and denote

$$\mu = \frac{m}{3} \int_0^\infty c\lambda_c \frac{dN_c}{dc} dc = \frac{1}{3} \varrho \langle c\lambda_c \rangle = \varrho v \tag{6}$$

if ϱ = density; m = molecular mass; v = kinematic viscosity;

$$\sigma^* = \frac{m}{3} \int_0^\infty l_c^2 \frac{dN_c}{dc} dc = \frac{1}{3} \varrho \langle l_c^2 \rangle = \varrho \sigma \tag{7}$$

one obtains the momentum flux density along x_1, and therefore the tangential stress $\tau_{x_1 x_2}^{(1)}$ (defined as the force per unit area that the fluid particles on the side of the x_2 positive axis exerts on those on the other side) which for the case under consideration is expressed by the rel.

$$\tau_{x_1 x_2}^{(1)} = \mu \frac{\partial u_1}{\partial x_2} - \sigma^* \frac{\partial^2 u_1}{\partial x_2 \partial t}. \tag{8}$$

Assume now that the gas flow has a velocity in the direction of the x_2 axis and function only of the coordinate x_1 i.e. $u = u_2(x_1, t) \, \boldsymbol{i}_2$, and follow a similar procedure as before to calculate the momentum transferred in the direction of the x_1 axis through the element dS by the thermal molecular motion. The fluid element, to which the molecules crossing dS belong, rotates with the velocity $\omega = \dfrac{\partial u_2}{\partial x_1}$ round the x_3 axis; at the moment t_0 the couple of slant cylinders having the basis dS and whose axes lying in the plane (x_1, x_2) are inclined of the angle θ to the positive x_2 axis, has the configuration indicated in Fig. 2.

Let be α the angle that the plane (r) of the basis dS makes with the (x_1, x_3) plane in such a moment.

If $t_1 = 0$ is the moment at which (r) in its rotating motion get lying in the (x_1, x_3) plane, it is $t_0 = -l/c$ and

$$\alpha = -\int_{-l/c}^0 \omega \, dt = \left(\frac{\partial u_2}{\partial x_1} \right)_0 \frac{l}{c} - \left(\frac{\partial^2 u_2}{\partial x_1 \partial t} \right)_0 \left(\frac{l}{c} \right)^2 \tag{9}$$

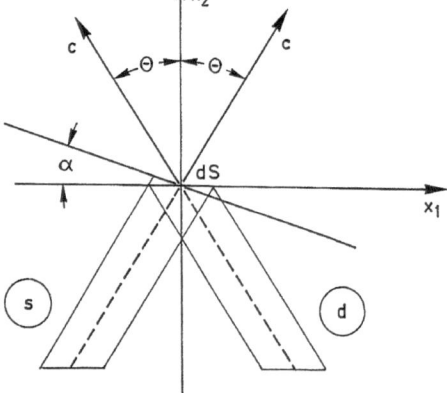

Fig. 2. Transfer of momentum in case II: $\boldsymbol{u} = u_2(x_1, t) \, \boldsymbol{i}_2$

where the suffix *zero* indicates that the values of the derivatives are calculated for $t = 0$. Now, anyone of the molecules contained in the slant cylinder d crossing dS with the velocity $(c; \theta)$ transports, during the interval of time dt, in the half-plane $x_2 > 0$ a momentum whose component along x_1 is equal to

$$mc_t \, dt = mc \cos(\pi/2 + \alpha + \theta) \, dt \simeq -mc \sin \theta \, dt - mc\alpha \cos \theta \, dt. \tag{10}$$

On the other hand, for a correspondent molecule into cylinder s it is

$$mc_t \, dt \simeq (mc \sin \theta - mc\alpha \cos \theta) \, dt \tag{11}$$

so that for the couple of these slant cylinders the transport of the component of momentum along x_1 for molecule turns out to be

$$-mc\alpha \cos \theta \, dt = -m \cos \theta l \left(\frac{\partial u_2}{\partial x_1} - \frac{\partial^2 u_2}{\partial x_1 \, \partial t} \frac{l}{c} \right) dt \tag{12}$$

and therefore for all the molecules $dn_{c\theta}$

$$-m \, dn_{c\theta} \cos \theta \left(\lambda_c \frac{\partial u_2}{\partial x_1} - \frac{l_c^2}{c} \frac{\partial^2 u_2}{\partial x_1 \, \partial t} \right) dt = -m \frac{c}{2} \sin \theta \cos^2 \theta \, d\theta \, dt \, dN_c \left(\lambda_c \frac{\partial u_2}{\partial x_1} - \frac{l_c^2}{c} \frac{\partial^2 u_2}{\partial x_1 \, \partial t} \right). \tag{13}$$

By integrating with respect to θ for θ ranging between zero and $\pi/2$ one obtains

$$-\frac{m}{6} \left(\lambda_c c \frac{\partial u_2}{\partial x_1} - l_c^2 \frac{\partial^2 u_2}{\partial x_1 \, \partial t} \right) dN_c. \tag{14}$$

A similar expression, but with opposite sign, one derives for the molecules crossing dS in the negative direction of the x_2 axis, so that the flux of momentum along x_1 results to be

$$-m \, dN_c \frac{1}{3} \left(c\lambda_c \frac{\partial u_2}{\partial x_1} - l_c^2 \frac{\partial^2 u_2}{\partial x_1 \, \partial t} \right) \tag{15}$$

and integrating with respect to c between zero and infinity one obtains for $\tau_{x_1 x_2}^{(2)}$ defined as above,

$$\tau_{x_1 x_2}^{(2)} = \mu \frac{\partial u_2}{\partial x_1} - \sigma^* \frac{\partial^2 u_2}{\partial x_1 \, \partial t}. \tag{16}$$

Consequently, if the velocity of the gas is

$$u = u_1(x, t) \, \boldsymbol{i}_1 + u_2(x, t) \, \boldsymbol{i}_2 \tag{17}$$

the tangential stress is given by

$$\tau_{x_1 x_2} = \mu \left(\frac{\partial u_2}{\partial x_1} + \frac{\partial u_1}{\partial x_2} \right) - \sigma^* \left(\frac{\partial^2 u_2}{\partial t \, \partial x_1} + \frac{\partial^2 u_1}{\partial t \, \partial x_2} \right). \tag{18}$$

Consider then the case in which it is $u = u_1(x_1, t) \, \boldsymbol{i}_1$ and calculate the flux of momentum along the x_1 axis trough the element dS lying in the plane (x_2, x_3); following a procedure similar to that used in the first of the cases above considered one obtains

$$\tau_{x_1 x_1}^{(1)} = \mu \frac{\partial u_1}{\partial x_1} - \sigma^* \frac{\partial^2 u_1}{\partial x_1 \, \partial t}; \tag{19}$$

but, to this normal stress it is still necessary to add two other contributions: the first is obtained taking in account the relative velocity to dS of any molecule crossing the element in the direction of positive x_1 which is given by the rel.:

$$c i_\theta - \left[\left(\frac{\partial u_1}{\partial x_1} \right) l \cos \theta - \left(\frac{\partial^2 u_1}{\partial x_1 \partial t} \right) \frac{l^2}{c} \cos \theta \right] i_1 . \tag{20}$$

As a consequence the number $dn_{c\theta}$ of these molecules is now equal to

$$dn_{c\theta} = \frac{c^2}{2} \sin \theta \cos \theta \, dt \left[1 - \left(\frac{\partial u_1}{\partial x_1} \right)_0 \frac{l}{c} + \left(\frac{\partial^2 u_1}{\partial x_1 \partial t} \right)_0 \frac{l^2}{c^2} \right] dN_c . \tag{21}$$

Since the same expression, with the opposite sign for the terms containing the derivative with respect to x_1 is valid for the molecules crossing dS in the negative direction of x_1 one obtains

$$\tau^{(2)}_{x_1 x_1} = \mu \frac{\partial u_1}{\partial x_1} - \sigma^* \frac{\partial^2 u_1}{\partial x_1 \partial t} . \tag{22}$$

The second further contribution follows from the difference of the values of the number dN_c of the molecules in the slant cylinder in either side of the plane (x_2, x_3) as a consequence of the variation of the velocity with x_1.

In the case now under consideration, being the density ϱ equal to mN, if N is the number of density of the molecules, it is $\dfrac{dN}{dt} = -N \dfrac{\partial u_1}{\partial x_1}$; therefore, if N_s is the value of N in the cylinder on the side of x_1 negative, N_0 the value of N when $x = 0$ and the moment $t_0 = t_1 - \dfrac{l}{c}$, it is

$$N_s - N_0 = \Delta N = -N_0 \left[\left(\frac{\partial u_1}{\partial x_1} \right)_0 - \left(\frac{\partial^2 u_1}{\partial x_1 \partial t} \right)_0 \frac{l}{c} \right] \frac{l}{c} \tag{23}$$

while the variation of the number of density in the cylinder on the other side is given by the same relation but with the opposite sign.

One can therefore write

$$\tau^{(3)}_{x_1 x_1} = \beta \left(\mu \frac{\partial u_1}{\partial x_1} - \sigma^* \frac{\partial^2 u_1}{\partial x_1 \partial t} \right) . \tag{24}$$

The relations above indicated give the irreversible transfer of momentum from points where the velocity is larger to those where it is smaller; to this flux one has to add the term corresponding to the pressure $P = Nm\langle c^2 \rangle$, so that the final expression of the normal stress is

$$\tau_{x_1 x_1} = (2 + \beta) \left(\mu \frac{\partial u_1}{\partial x_1} - \sigma^* \frac{\partial^2 u_1}{\partial x_1 \partial t} \right) - P \tag{25}$$

and if one accepts the Stokes hypothesis $2 + \beta = 4/3$

$$\tau_{x_1 x_1} = \frac{4}{3} \left(\mu \frac{\partial u_1}{\partial x_1} - \sigma^* \frac{\partial^2 u_1}{\partial x_1 \partial t} \right) - P . \tag{26}$$

By applying the same procedure again and again one determine all the other components of the stress tensor, whose expression can be obtained by that given by the Rel. (26) with a simple permutation of the symbols, so that one can write

$$\tau_{ik} = \mu \left(\gamma_{ik} - \frac{2}{3} \delta_{ik} \frac{\partial u_l}{\partial x_l} \right) - \sigma^* \left(\frac{\partial}{\partial t} \gamma_{ik} - \frac{2}{3} \delta_{ik} \frac{\partial^2 u_l}{\partial x_l \partial t} \right) - P\delta_{ik}$$

$$= \mu \Gamma_{ik} - \sigma^* \frac{\partial}{\partial t} \Gamma_{ik} - P\delta_{ik} = -P\delta_{ik} + \tau_{ik}^* \tag{27}$$

denoting

$$\gamma_{ik} = \frac{\partial u_i}{\partial x_k} + \frac{\partial u_k}{\partial x_i}; \quad \Gamma_{ik} = \gamma_{ik} - \frac{2}{3} \delta_{ik} \frac{\partial u_l}{\partial x_l};$$

$$\delta_{ik} = \begin{cases} 0 & \text{if } i \neq k \\ 1 & \text{if } i = k. \end{cases} \tag{28}$$

It is still suitable to see that the same results can be obtained taking the distribution function of the molecular velocities as that assumed by A. Lorentz [8] and M. M. Brillouin [5] for gas in state of quasi-equilibrium, and following the method by them indicated. It is

$$f(u_1, u_2, u_3) = \left(\frac{hm}{\pi} \right)^{1/2} \exp \left[-hm(u_1'^2 + u_2'^2 + u_3'^2) (1 + \varepsilon) \right]. \tag{29}$$

Where $u = u_1' + u_0$, being u_0 the mean value of u, while ε can be expressed by a series of powers of u' and it is invariant in form and in value with respect of any change of orthogonal axes. Following the method used by the Authors above quoted (Y. Rocard [6]), ε, apart from the terms depending on the temperature, is considered a linear function of the following invariants

$$c^2 = u' \cdot u'; \quad \nabla \cdot u_0;$$

$$u' \cdot \left[u' \cdot \left(\nabla u_0 - t^* \frac{\partial}{\partial t} \nabla u_0 \right) \right] \tag{30}$$

where ∇ is the operator $\nabla = i_k \frac{\partial}{\partial x_k}$ $(k = 1, 2, 3)$; t^* is a characteristic time whose expression will be defined below.

One deduces that the stress tensor has the expression:

$$\tau_{ik} = -P\delta_{ik} + \mu \left(\gamma_{ik} - \frac{2}{3} \delta_{ik} \frac{\partial u_l}{\partial x_l} \right) - \mu t^* \left(\frac{\partial}{\partial t} \gamma_{ik} - \frac{2}{3} \delta_{ik} \frac{\partial}{\partial t} \frac{\partial u_l}{\partial x_l} \right). \tag{31}$$

By comparison of the Rels. (31) and (27) one obtains $t^* = \sigma^*/\mu$. From the expression of σ^* and μ given by (6) and (7) one derives

$$t^* = \frac{\langle l_c^2 \rangle}{\langle c\lambda_c \rangle}, \tag{32}$$

and since the order of magnitude of $\langle l_c^2 \rangle$ may be considered equal to that of $\langle \lambda_c \rangle^2$ one has

$$t^* = O \left(\frac{\langle \lambda_c \rangle}{c} \right). \tag{33}$$

Now, from the equation defining the variation law of the entropy it appears that, apart the terms depending on the temperature, the variation of the entropy due to the viscosity is given by the dissipation function of the energy, that independently from the form of the constitutive equation, is given by

$$\Phi = (\tau_{ik} - \delta_{ik}P)\frac{\partial u_{0i}}{\partial x_k}$$

and therefore, now

$$\Phi = \mu\left[\left(\gamma_{ik} - \frac{2}{3}\delta_{jk}\frac{\partial u_{0l}}{\partial x_l}\right) - t^*\frac{\partial}{\partial t}\left(\gamma_{ik} - \frac{2}{3}\delta_{ik}\frac{\partial u_{0l}}{\partial x_l}\right)\right]\frac{\partial u_{0i}}{\partial x_k}. \tag{34}$$

It is then

$$\mu\left(\gamma_{ik} - \frac{2}{3}\delta_{ik}\frac{\partial u_{0l}}{\partial x_l}\right)\frac{\partial u_{0i}}{\partial x_k} = \frac{1}{2}\mu\left(\gamma_{ik} - \frac{2}{3}\delta_{ik}\frac{\partial u_{0l}}{\partial x_l}\right)^2 \tag{35}$$

while, denoting by t_c a characteristic time during which the quantity that is differentiated with respect to time and it is indicated between brackets shows a notable variation, and taking $t = t_c T$ one obtains

$$\mu t^*\frac{\partial}{\partial t}\left[\left(\gamma_{ik} - \frac{2}{3}\delta_{ik}\frac{\partial u_{0l}}{\partial x_l}\right)\right]\frac{\partial u_{0i}}{\partial x_k} = \frac{1}{2}\mu\frac{t^*}{t_c}\frac{\partial}{\partial T}\left[\gamma_{ik} - \frac{2}{3}\delta_{ik}\frac{\partial u_{0l}}{\partial x_l}\right]^2. \tag{36}$$

In order that Φ be positive for whichever $u_0(x, t)$ it is necessary and sufficient that be positive all the terms in it contained, and since

$$\frac{\dfrac{\partial}{\partial T}\left(\gamma_{ik} - \dfrac{2}{3}\dfrac{\partial u_{0l}}{\partial x_l}\right)^2}{\left(\gamma_{ik} - \dfrac{2}{3}\dfrac{\partial u_{0l}}{\partial x_l}\right)^2} \equiv O(1) \tag{37}$$

it must be $\dfrac{t^*}{t_c} \ll 1$, i.e. $t_c \gg \dfrac{\langle \lambda_c \rangle}{c}$.

This condition may be considered equivalent to the condition that the fluid is continuous.

3 The motion equations

It is assumed here that μ and σ^* are constant; take in account their variation with temperature and pressure makes the equations of motion more complicated, without, however, changing their properties.

From the Eq. (27) one obtains

$$\frac{\partial}{\partial t}\tau_{ik}^* = \mu\frac{\partial\Gamma_{ik}}{\partial t} - \sigma^*\frac{\partial^2\Gamma_{ik}}{\partial t^2} \tag{38}$$

and therefore

$$\tau_{ik}^* = \mu\Gamma_{ik} - \sigma^*\left(\frac{1}{\mu}\frac{\partial}{\partial t}\tau_{ik}^* + \frac{\sigma^*}{\mu}\frac{\partial^2}{\partial t^2}\Gamma_{ik}\right). \tag{39}$$

But it is

$$\frac{\sigma^* \dfrac{\partial^2}{\partial t^2} \Gamma_{ik}}{\dfrac{\partial}{\partial t} \tau_{ik}^*} \equiv O\left(\frac{\sigma^*/t_c{}^2}{\mu/t_c}\right) \equiv \left(\frac{t^*}{t_c}\right) \ll 1 \tag{40}$$

so that one can take

$$\tau_{ik}^* = \mu \Gamma_{ik} - \frac{\sigma^*}{\mu} \frac{\partial}{\partial t} \tau_{ik}^*. \tag{41}$$

On the other hand, by the motion equations

$$\varrho \frac{\partial u_i}{\partial t} + \varrho u_k \frac{\partial u_i}{\partial x_k} = -\frac{\partial P}{\partial x_i} + \frac{\partial}{\partial x_k} \tau_{ik}^* \tag{42}$$

one deduces by differentiating with respect to t

$$\frac{\partial}{\partial t}\left(\varrho \frac{\partial u_i}{\partial t}\right) + \frac{\partial}{\partial t}\left(\varrho u_k \frac{\partial u_i}{\partial x_k}\right) = -\frac{\partial^2 P}{\partial x_i \, \partial t} + \frac{\partial^2}{\partial t \, \partial x_k} \tau_{ik}^* \tag{43}$$

and since it is

$$\sigma^* \frac{\partial^2}{\partial t \, \partial x_k} \tau_{ik}^* = \mu^2 \frac{\partial \Gamma_{ik}}{\partial x_k} - \mu \frac{\partial \tau_{ik}^*}{\partial x_k} \tag{44}$$

it results

$$\sigma^* \left[\frac{\partial}{\partial t}\left(\varrho \frac{\partial u_i}{\partial t}\right) + \frac{\partial}{\partial t}\left(\varrho u_k \frac{\partial u_i}{\partial x_k} + \frac{\partial P}{\partial x_i}\right)\right] = \mu^2 \frac{\partial \Gamma_{ik}}{\partial x_k} - \mu \frac{\partial \tau_{ik}^*}{\partial x_k}. \tag{45}$$

There, one can write

$$\frac{\partial \Gamma_{ik}}{\partial x_k} = \frac{\partial^2 u_i}{\partial x_k \, \partial x_k} + \frac{1}{3}\frac{\partial}{\partial x_i}\frac{\partial u_l}{\partial x_l};$$
$$\varrho \frac{du_i}{dt} + \frac{\partial P}{\partial x_i} = \frac{\partial}{\partial x_k} \tau_{ik}^* \tag{46}$$

and therefore

$$\sigma^* \left[\frac{\partial}{\partial t}\left(\varrho \frac{du_i}{dt}\right) + \frac{\partial P}{\partial x_i}\right] = -\mu \varrho \frac{du_i}{dt} - \mu \frac{\partial P}{\partial x_i} + \mu^2 \left(\frac{\partial^2 u_i}{\partial x_k \, \partial x_k} + \frac{1}{3}\frac{\partial}{\partial x_i}\frac{\partial u_l}{\partial x_l}\right) \tag{47}$$

that when $\sigma^* = 0$ becomes the classical Navier-Stokes equations.

4 Plane waves

Let us apply the equations above indicated to the case of slow non stationary motion in the direction of the axis $x \equiv x_1$, and therefore it is $u_2 = u_3 = 0$, $u_1 = u(x, t)$.

Eq. (47) becomes

$$\sigma^* \left(\varrho \frac{\partial^2 u}{\partial t^2} + \frac{\partial^2 P}{\partial x \, \partial t}\right) - \frac{4}{3} \mu^2 \frac{\partial^2 u}{\partial x^2} + \mu \varrho \frac{\partial u}{\partial t} + \mu \frac{\partial P}{\partial x} = 0. \tag{48}$$

To the Eq. (48) one has to add the continuity equation

$$\frac{\partial \varrho}{\partial t} + \varrho \frac{\partial u}{\partial x} = 0,$$ (49)

that can be written under the form

$$\frac{\partial P}{\partial t} + \varrho c_s^2 \frac{\partial u}{\partial x} = 0$$ (50)

where $c_s = \left(\frac{dP}{d\varrho}\right)_{s=\text{const}}$ if s is the entropy of the fluid, and therefore $c_s =$ the sound velocity in a non-viscous fluid. Let be

$$u = Uc_s; \quad x = LX; \quad t = \frac{L}{c_s} T; \quad P = \varrho c_s^2 P; \quad L = \frac{v}{u_0}$$ (51)

being u_0, now, a reference value of the velocity, that characterizes the order of magnitude of it. The Eqs. (48), (50) become

$$\frac{\partial^2 U}{\partial T^2} - K \frac{\partial^2 U}{\partial X^2} + h \frac{\partial U}{\partial T} + h \frac{\partial P}{\partial X} = 0; \quad \frac{\partial P}{\partial T} + \frac{\partial U}{\partial X} = 0$$ (52)

where the meaning of the constants K and h is the following:

$$K = 1 + \frac{4}{3} \frac{\mu^2}{\sigma^* \varrho c_s^2} = 1 + \frac{4}{3} \frac{v^2}{\sigma c_s^2}; \quad h = \frac{vL}{\sigma c_s} = \frac{v^2}{\sigma u_0 c_s}.$$ (53)

The system (52) can be easily reduced to only one differential equation of the third order by eliminating P and its solution can be obtained applying the classical methods.

It seems, however, that the behaviour of the fluid in the phenomenon under consideration can be pointed out more easily and clearly following the procedure below indicated. Denoting

$$\frac{\partial U}{\partial T} = p; \quad \frac{\partial U}{\partial X} = q$$

the system (52) can be written in the following way

$$\frac{\partial p}{\partial T} - K \frac{\partial q}{\partial X} + hp + h \frac{\partial P}{\partial X} = 0;$$

$$\frac{\partial q}{\partial T} - \frac{\partial p}{\partial X} = 0;$$ (54)

$$\frac{\partial P}{\partial T} + q = 0.$$

In matrix notation one can write

$$L(V) = A V_X + B V_T = C; \quad V \equiv (p, q, T)$$ (55)

being the matrices A and B given by

$$A = \begin{vmatrix} 0 & -K & h \\ -1 & 0 & 0 \\ 0 & 0 & 0 \end{vmatrix};$$

$$B = \begin{vmatrix} 1 & 0 & 0 \\ 0 & 1 & 0 \\ 0 & 0 & 0 \end{vmatrix} = I.$$

(56)

While the vector C is defined by the relations

$$C^1 = -hp; \quad C^2 = 0; \quad C^3 = -q.$$

(57)

The characteristic lines of the system (54) are defined by the equations $\dfrac{dX}{dT} = \tau$ where τ are the eigenvalues of the matrix A and therefore the roots of the characteristic equation

$$|A - \tau I| = 0 = \begin{vmatrix} -\tau & -K & h \\ -1 & -\tau & 0 \\ 0 & 0 & -\tau \end{vmatrix}.$$

(58)

One deduce at once $-\tau^3 + \tau K = 0$ from which it derives

$$\tau_1 = 0; \quad \tau_2 = K^{1/2}; \quad \tau_3 = -K^{1/2}.$$

(59)

The correspondent eigenvectors are deduced from the eqn. $l \cdot A = \tau l$ and therefore one obtains

for $\tau_1 = 0$:
$$l_1 0 - l_2 + l_3 0 = 0;$$
$$-Kl_1 + l_2 0 + l_3 0 = 0;$$
$$l_1 h + l_2 0 + l_3 0 = 0$$

(60)

from which $l_1 = l_2 = 0$; l_3 arbitrary

for $\tau_2 = K^{1/2}$:
$$-\tau l_1 - l_2 = 0;$$
$$-l_1 K - l_2 \tau = 0;$$
$$l_1 h - \tau l_3 = 0$$

(61)

from which:
$$l_1 = -l_2 K^{-1/2};$$
$$l_3 = l_1 h K^{-1/2} = -l_2 h K^{-1}.$$

(62)

for $\tau_3 = -K^{1/2}$:
$$l_1 = l_2 K^{-1/2};$$
$$l_3 = -l_1 h K^{-1/2}.$$

(63)

The characteristic lines are therefore the straightlines

$$\xi = X - K^{1/2} T; \quad \eta = X + K^{1/2} T; \quad \zeta = X.$$

(64)

The propagation velocity of the correspondent plane waves results to be $\dfrac{dX}{dT} = \pm K^{1/2}$ or in the physical plane (x, t): $\dfrac{dx}{dt} = \pm K^{1/2} c_s$.

One obtains then the variation law of the velocity and pressure along the characteristic lines from the eqn.:

$$l_i(AV_X + IV_T) = l_i C^i \tag{65}$$

and therefore: along the characteristic lines $\xi = \text{const.}$:

$$l_1\left(\frac{\partial p}{lT} + K^{1/2}\frac{\partial p}{\partial X}\right) + l_2\left(\frac{\partial q}{\partial T} + K^{1/2}\frac{\partial q}{\partial X}\right) + l_3\left(\frac{\partial P}{\partial T} + K^{1/2}\frac{\partial P}{\partial X}\right) = -hl_1 p - l_3 q. \tag{66}$$

By substitution of the values of the components l_1 above given, and observing that it is

$$\frac{\partial}{\partial\eta} = \frac{1}{2K^{1/2}}\left(\frac{\partial}{\partial T} + K^{1/2}\frac{\partial}{\partial X}\right) \tag{67}$$

one deduces

$$\frac{\partial}{\partial\eta}(K^{1/2}p - Kq + hP) = -\frac{h}{2}K^{-1/2}(q + K^{1/2}p). \tag{68}$$

Along the characteristic $\eta = \text{const}$, using the correspondent values of l_i and following a similar procedure as before, one obtains

$$\frac{\partial}{\partial\xi}(K^{1/2}p + Kq - hP) = \frac{h}{2}K^{-1/2}(K^{1/2}p - q) \tag{69}$$

and finally, for $\zeta = \text{const}$ it is

$$\frac{\partial P}{\partial T} = -q. \tag{70}$$

The equations now obtained are very suitable to the numerical solution in any initial value problem: given the initial values of $U, \dfrac{\partial U}{\partial T}$ and P in the range $(0 \div a)$ of the X axis (and therefore the values of p, q, P), to determine these functions in the domain in which they are defined (Cauchy problem). For any point M in the plane (X, T) the domain of dependence of M results to

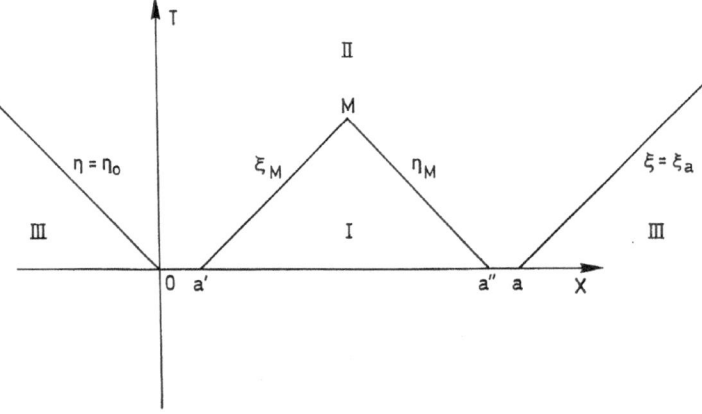

Fig. 3. Domain of dependence (I); Domain of influence (II); Silent regions (III$-$III$'$)

be the segment $\Sigma(a'a'')$ of X intercepted on said axis by the two characteristics of slope τ_2 and τ_3 (Fig. 3), while the domain of influence of the data on the segment $(0, a)$ is the region of the half-plane $T > 0$ lying between the two characteristics $\eta = \eta_0$; $\xi = \xi_a$ pointing from the ends points of Σ into the upper half-plane. It turns out that in the regions III and III' it is not felt, whichever be X, the disturbancies produced in $(0 \div a)$.

The slope of the two characteristics is dependent on the value of the constant K: if $\sigma = 0$ (and therefore the motion is determined by the Navier-Stokes equations) it is $K = \infty$, the domain of dependence in the point M is all the X axis, and the domain of influence is all the half-plane $(X, T > 0)$; the disturbancies produced in the points of the segment $(0 \div a)$ propagate in all the half-plane and the velocity of propagation is infinite. Now, it is notworthy that, notwhitstanding that the additional term in the constitutive equation is very small with respect to the classical one, the parameter K is very near to *one*: in fact, for a gas like air in ordinary conditions of temperature and pressure it is $v = 0.132$ cm/sec.

$$c_s = 3.4 \times 10^4 \text{ cm/sec.}; \quad \sigma = 2.9 \times (2 \times 10^{-6})^2 \text{ cm}^2; \tag{71}$$

from which it results

$$K^{1/2} = 1.045. \tag{72}$$

5 Solution of the Cauchy problem in the flow by plane-waves

Let be

$$V_1 = K^{1/2}p + Kq - hP;$$

$$V_2 = K^{1/2}p - Kq + hP; \tag{73}$$

$$V_3 = P.$$

Solving these relations with respect to $K^{1/2}p$; Kq; hP one obtains

$$K^{1/2}p = \frac{1}{2}(V_1 + V_2);$$

$$Kq = \frac{1}{2}(V_1 - V_2) + hV_3; \tag{74}$$

$$hP = hV_3$$

and therefore

$$K^{1/2}p - q = \frac{1}{2}\frac{K-1}{K}V_1 + \frac{1}{2}\frac{K+1}{K}V_2 - \frac{h}{K}V_3;$$

$$q + K^{1/2}p = \frac{1}{2}\frac{K+1}{K}V_1 + \frac{1}{2}\frac{K-1}{K}V_2 + \frac{h}{K}V_3. \tag{75}$$

The Eqs. (68), (69), (70) become

$$\frac{\partial}{\partial \xi} V_1 = \frac{h}{2} K^{-1/2} \left(\frac{1}{2} \frac{K-1}{K} V_1 + \frac{K+1}{2K} V_2 - \frac{h}{K} V_3 \right) = F_1(V_1, V_2, V_3);$$

$$\frac{\partial}{\partial \eta} V_2 = -\frac{h}{2} K^{-1/2} \left(\frac{1}{2} \frac{K+1}{K} V_1 + \frac{1}{2} \frac{K-1}{K} V_2 - \frac{h}{K} V_3 \right) = F_2(V_1, V_2, V_3); \quad (76)$$

$$\frac{\partial}{\partial T} V_3 = -\frac{1}{K} \left[\frac{1}{2} (V_1 - V_2) + hV_3 \right] = F_3(V_1, V_2, V_3).$$

The Eq. (25) has the same form as that indicated by R. Courant [6] to construct the solution of the problem by an iterative method: by integrating each of the Eqs. (76) along the corresponding characteristic from the point P_i intersection of this characteristic with the X axis, to the point $P(\xi, \eta)$ (Fig. 4) one obtains

$$V_1(P) = V_1(P_1) + \int_{P_1}^{P} F_1(V_1, V_2, V_3) \, d\xi. \quad (77)$$

While the other quantities V_i are given by similar expression. Starting from admissible functions $V_i^{(0)}$ that in the points P_i of the X axis take the prefixed values $V_i(P_i)$ one constructs by iteration the functions V_i as the limits for $n \to \infty$ of the equations

$$V_1^{(n+1)} = V_i(P_i) + \int_{P_1}^{P} F_i(V_1^{(n)}, V_2^{(n)}, V_3^{(n)}) \, d\xi_i; \quad (78)$$

$$(\xi_1 = \xi; \xi_2 = \eta; \xi_3 = T).$$

Now, when $K = 1$ the solution of the problem can be easily obtained: it is in fact, in this case

$$K^{1/2} p - q = \frac{\partial U}{\partial T} - \frac{\partial U}{\partial X} = -2 \frac{\partial U}{\partial \xi};$$

$$q + K^{1/2} p = 2 \frac{\partial U}{\partial \eta} \quad (79)$$

so that the first two Eqs. (76) become

$$\frac{\partial}{\partial \xi} (V_1 + hU) = 0; \quad \frac{\partial}{\partial \eta} (V_2 + hU) = 0. \quad (80)$$

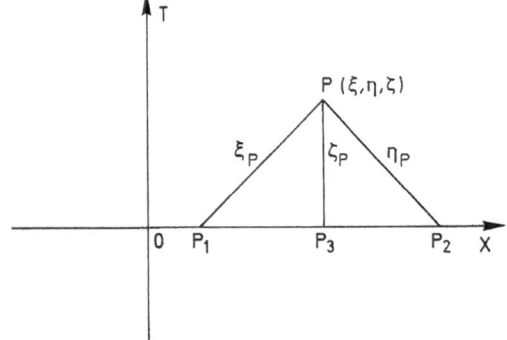

Fig. 4. Scheme for the application of the iteration method

It turns out that

$$V_1 + hU = 2f_1(\eta); \qquad V_2 + hU = 2f_2(\xi) \tag{81}$$

from which one obtains

$$V_1 + V_2 + 2hU = 2[f_1(\eta) + f_2(\xi)]. \tag{82}$$

Since $V_1 + V_2 = 2p = 2\dfrac{\partial U}{\partial T}$ it is

$$\frac{\partial U}{\partial T} + hU = f_1(\eta) + f_2(\xi). \tag{83}$$

Beside

$$V_1 - V_2 = 2q - 2hP = 2[f_1(\eta) - f_2(\xi)] = -2\frac{\partial P}{\partial T} - 2hP. \tag{84}$$

From the Eq. (84) one deduces

$$P = P_0(X, 0) \exp(-hT) + \int_0^T \exp[-h(T - T_1)] \cdot [f_1(X + T_1) - f_2(X - T_1)]\, dT_1 = V_3^{(0)} \tag{85}$$

while one obtains from the Eq. (27)

$$U = U_0(X, 0) \exp(-hT) + \int_0^T \exp[-h(T - T_1)] [f_1(X + T_1) + f_2(X - T_1)]\, dT_1 = U^{(0)};$$

$$\begin{aligned}
p = \frac{\partial U}{\partial T} &= -hU_0(X, 0) \exp(-hT) + f_2(\xi) + f_1(\eta) \\
&\quad - h \int_0^T \exp[-h(T - T_1)]\,(f_1 + f_2)\, dT_1 = p^{(0)};
\end{aligned} \tag{86}$$

$$q = \frac{\partial U}{\partial X} = \frac{\partial U_0(X, 0)}{\partial X} \exp(-hT) + \int_0^T \exp[-h(T - T_1)] \left(\frac{\partial f_1}{\partial X} + \frac{\partial f_2}{\partial X}\right) dT_1 = q^{(0)}.$$

From the Eqs. (86) and from the initial data one obtains the values of $V_1^{(0)}$, $V_2^{(0)}$, $V_3^{(0)}$, that allows to start for the iterative procedure.

6 Conclusions

The simple case above considered shows that if one takes in account the dependence on the time of the distribution function of the molecular velocities one deduces by the corresponding motion equations that the propagation of small perturbations occurs really by means of waves in similar way as in the case of non viscous fluid. The phenomenon is governed by two parameters: K, which is linked essentially to the velocity propagation of the waves, and it is always greater than one, so that this velocity is always greater than that in non viscous fluid; h, from which is depending the attenuation of the perturbations. The first of these results seems to be in qualitative agreement with the representation given by C. Possio, for which the median line of the propagation layer moves with the sound speed while the thickness of said layer increases with the time. As far as it concerns the comparison with the case of non viscous fluid, the mean difference is not in the different values of the wave velocity, since $(K - 1) \ll 1$, but in the property that in non viscous fluid the characteristic lines are discontinuity lines of the first derivatives of the fluid velocity, while in viscous fluid the discontinuity is in the second derivates of said velocity.

References

[1] Duhem, P.: Récherches sur l'élasticité. Paris: Gauthier-Villars 1906.

[2] Levi-Civita, T.: Caractéristiques des Systèmes différentiels et Propagation des ondes. Paris: Librairie Félix Alcan 1932.

[3] Possio, C.: L'influenza della viscosita' e della conducibilita' termica sulla propagazione del suono R. Accademia delle Scienze di Torino, Vol. 78. 1942.

[4] Cattaneo, C.: Seminario Matematico e Fisico dell'Università di Modena. Modena: Società Tipografica Modenese 1948.

[5] Brillouin, M. M.: Théorie Moleculaire des Gas. Diffusion de Mouvement et de l'Energie. Annales de Chémie et de Physique. Vol. 20 (XXV). 1900.

[6] Rocard, Y.: L'Hydrodynamique et la Théorie cinétique des gas. Paris: Gauthier-Villars 1932.

[7] Courant, R.: Partial differential equations. New York—London: Interscience Publishers, John Wiley and Sons 1962.

[8] Lorentz, H. A.: Annalen der Physik. Vol. 12. 1881.

Authors' address: Prof. Carlo Ferrari, Politecnico di Torino, Corso Galileo Ferraris 146, I-10129 Torino, Italy.

Acta Mechanica (1992) [Suppl] 3: 83–98

Wave propagation due to pneumatic surge in diabatic circuits and systems

A. **Romiti**, T. **Raparelli**, and E. **Bertello**, Torino, Italy

Summary. This study describes the behaviour of circuits with a pneumatic or otherwise gaseous working or control fluid during fast transients. The fluid energy evolution mode is defined taking into account friction and thermal exchange through the walls. Given the short times involved and the large thermal capacity of the walls by comparison with fluid mass, wall temperature is assumed to be constant. Wave propagation is supposed to occur only in continuous parameters (ducts), while vessels, actuators, restrictors, branchings and so on are considered lumped. Wave propagation equations are hyperbolic and can be reduced to ordinary differential equations along characteristic lines. Dependent variables are pressure, volumetric flow per unit time and density (or temperature). Boundary conditions of the characteristic equations are expressed by the relations amongst dependent variables at duct input and output, which are given by analysis of the joined components. A practical procedure for numerical analysis of pneumatic or gaseous circuits with any number of components and branches has been developed from this theoretical basis, and has been shown to provide very good agreement with experimental results. Furthermore, it appears that the method developed herein is the only one capable of simulating the behaviour of fast circuit transient closely.

1 Introduction

Scope of this paper is to present a computerized method for finding all the characteristic quantities related to pneumatic systems during short transients. The system may consist of any number of components, including valves, vessels, actuators, and restrictors connected by ducts of any length. The quantities to be found are pressure, density, temperature and mean velocity of the fluid in any circuit section and position, speed and acceleration of solid moving parts (if any). All quantities are functions of time and of location. The adopted method relies on the construction of a model of the system, through identification of the significant characteristics of all components. This is followed by digital simulation, starting from the initial conditions that produce the transient process. Digital simulation appears to be an extremely powerful design tool for fluid power engineers. It makes it possible to forecast system behaviour even under extreme conditions.

Much effort has been devoted to studying hydraulic circuits, and a number of simulation packages are now available, such as, DSH [1], Hopsan [2], Catsim [3], Hycad [4] and Bath [5], while other simulation programs were presented in [6] and [7]. Less effort has been devoted to pneumatic circuits and systems. No studies have been made of complete pneumatic systems; rather, attention has been limitated to single elements such as ducts [8 – 10], ducts with branching [11] or actuators [12].

The studies from [8] to [11] are based on the method of characteristics applied to transient gas flow in ducts. Ref. [8] gives a very simple treatment of the problem, by solving at the start the thermal and density problems by assuming an isoentropic one-space dimensional flow and neglecting friction forces. The assumptions made in Refs. [9 − 11] admit an axisymmetric laminar flow, and their treatment of such assumptions is equivalent to considering one-space dimensional flow, as it is done in this paper. However, the following assumptions are also made in [9 − 11]:

(i) the duct walls are rigid (versus the assumption in this paper that the walls may be elastic, with an experimentally determined elasticity modulus);

(ii) the fluid is laminar (versus the assumption in this paper that it is turbulent, as it generally occurs in the case of pressure surge; the wall friction coefficients are in this paper evaluated experimentally for any type of duct, while they are obviously neglected in the laminar case);

(iii) temperature and compressibility effects are considered to be small; such assumption leads to a considerable simplification of the equation, versus the present treatment dealing with full consideration of such effects.

Reference [12] refers to a method for analyzing fluid transmission lines through infinite series representation, i.e., by an expansion of the linear method for small perturbation. Such a method appears to be unfit for implementing an efficient simulation for complex pneumatic system with several components, connecting ducts and with heat transfer.

Reference [13] covers the subject of heat transfer in vessels and actuator pneumatic chambers, and has been used in this work for obtaining important data and relationships on the subject. Reference [14] presents relevant heat transfer data.

From the above examination of the literature, it is apparent that no computerized system has yet been developed which is capable of accurately analyzing fast pneumatic transients in complex systems (including those involving a large number of components), or of dealing with large, rapid pressure changes. References [15] and [16] by the authors of the present paper, deal with their earlier works on the same subject that do not consider the influence of heat transfer. It was considered earlier ([15, 16]) that the walls of components and ducts containing the fluid in the system were adiabatic. In this case the heat transfer is obviously zero, and, even if the flow is not isoentropic (in fact friction effects were considered), a relatively simply modelling was possible. If the walls are diabatic, an approximate study of heat transfer is necessary even if the fluid temperature change is small; it is needed in any case to determine fluid thermodynamic transformation in the form of a relationship connecting fluid density with pressure, temperature and velocity. The present model has two main dependant variables, fluid pressure and flow throughput, which are functions of the variable time, and, in the case of the ducts, of the distance along the duct. Other dependant variables are density, temperature, and (where applicable) the speed of moving parts in actuators. The mathematical model is based on the equations of momentum, mass and energy conservation applied to fluid, on the equation of state for air and on the equation of motion applied to the moving parts. The simulation model does not consider effects of shock waves in air, that are noteworthy only for extremely short time intervals after an abrupt pressure surge at a given location, and vanish very quickly through repeated reflections and friction.

In this study, the following assumptions are made:

(i) the temperature of the walls of any component is considered constant and equal to the ambient temperature outside the circuit, because their heat capacity is far larger than for the fluid, and the fluid transit time is short.

(ii) expressions for heat transfer are taken from the literature.

(iii) shock wave effects are neglected, for the fastness of their vanishing time.

(iv) only the ducts are considered to be propagative elements. All other components are considered lumped.

(v) fluid flow in ducts is considered to be one-space dimensional.

The simulation approach is similar to that used for adiabatic circuits by the same authors [15], except for the greater complexity of the equations resulting from the need for continuous space/time processing of the energy equation.

2 System modelling

2.1 Component modelling

A component data base is first established. As indicated above, the components of a fluid system like valves, vessels, fluid resistors, and generally all components having a short length, i.e. a short run time of fluid perturbances through this length compared to a system characteristic time, are considered lumped, and thus not affected by propagative phenomena. Only the ducts interposed between two non-propagative components are considered to be involved in such propagative phenomena, i.e. there are time delays and gain variations along the duct length. Data about non-propagative components are given by instantaneous input-output relationships, such as flow throughput versus the ratio between the pressures at the two ends of the component. Such relationships are generally given by semi-empirical formulas, whose coefficients must be found experimentally. Futhermore, geometrical data such as the component inside volume and surface in contact with the fluid are necessary. In the case of components having solid moving parts, the mass and friction values of such parts are required, and force equilibrium including inertia terms give further relationships. Data about propagative components (ducts) include geometrical features like length and diameter, and physical features like duct wall expansion coefficient versus pressure and flow friction coefficient versus fluid speed. Data about fluid properties and characteristics, such as density, viscosity, thermal conductivity and specific heats, initial pressure and temperature, and also the outside temperature, are also necessary. Thermal characteristics need not to be specified only for those components that do not contain an appreciable fluid mass, where thermal exchange is not significant and the same stagnation temperature is maintained at inlet and outlet.

The method presented here may also be applied where the fluid is a liquid; in such cases all thermal characteristics may be neglected. For a gaseous fluid, if all the containing components walls are adiabatic, i.e., no heat transfer occurs through the walls, the only thermal characteristic of interest is the ratio of specific heats (at constant volume and at constant pressure). More details about component modelling are provided in Ref. [15], [16].

2.2 Thermal modelling

For modelling purposes, the heat transfer coefficient α from the fluid to the fluid containing walls of the components is given by Hausen's formula for the vessels and by Eckert-Drake's formula for the ducts [14]. Heat transfer through the walls of fluid resistances, such as valves or restrictors, is neglected because of the small fluid mass contained in these components.

For vessels, the following equation applies:

$$\frac{\alpha d}{K} = 0.116[(Re)^{2/3} - 125]\,(Pr)^{1/3}\left[1 + \left(\frac{d}{x}\right)^{2/3}\right]\left(\frac{\mu_B}{\mu_w}\right)^{0.14} \tag{1}$$

where α is the heat transfer coefficient from fluid to wall; d is the vessel diameter (or four times the ratio between inner area of a cross section and perimeter); K is a weak function of temperature (see K tables for air, e.g. Eckert and Drake [14]). Reynolds number Re and Prandtl number Pr are respectively defined by

$$Re = Vd/\nu_B$$

$$Pr = \mu_B c_v/\xi$$

where V is the inlet mean fluid speed; ξ is the fluid thermal conductivity; ν_B and μ_B are respectively the mean kinematic and dynamic fluid viscosity coefficient; μ_w is the dynamic viscosity coefficient of the fluid in contact with the vessel wall; c_v is the fluid specific heat at constant volume, and x is the distance of the considered section from the vessel inlet port.

According to [13], the various viscosity coefficients ν and μ are given by

$$\nu = [18.2 + (T - 298)\,0.044] \cdot 10^{-6} \tag{2}$$

$$\mu_B = \varrho_B \nu_B = \varrho_B[18.2 + (T - 298)\,0.044] \cdot 10^{-6} \tag{3}$$

$$\mu_w = \varrho_B \nu_w = \varrho_B[18.2 + (T_w - 298)\,0.044] \cdot 10^{-6} \tag{4}$$

where T is the mean absolute temperature in the vessel and T_w is the fluid temperature at the wall, which is considered equal to the ambient temperature outside the wall. The Prandtl number is considered approximately constant and equal to $Pr = 0.7$.

For ducts, on the other hand, the following equation applies:

$$\frac{\alpha}{\varrho c_p V} = \frac{0.038\,4Re^{-1/4}}{1 + 1.5Pr^{-1/6}(Pr - 1)\,Re^{-1/8}} \tag{5}$$

where ϱ and V are here, respectively, mean fluid density and velocity at a given duct section, c_p is the fluid specific heat at constant pressure.

Given α, (which is calculated by (1) or (5)), the heat flow ϕ per unit time flowing into the vessel or into the duct is

$$\phi = -\alpha S_w(T - T_w) \tag{6}$$

where S_w is the lateral surface of the vessel. The heat flow Φ per unit time and unit mass flowing into a duct is

$$\Phi = -\alpha(T - T_w)\,\frac{4}{\varrho D}. \tag{7}$$

In fact, if D is the duct diameter and ϱ the fluid density inside it, the mass of fluid contained by a surface S_w in a length l of tube is $\varrho(\pi D^2/4)\,l$ where $S_w = \pi Dl$; diving the heat flow per unit time given by (6) by the corresponding mass, gives (7).

3 The energy equation

Consider the volume W of a vessel or of an actuator chamber.

— For vessels, the energy equation may be written in the following form:

$$Q\left(\frac{p'}{\varrho'} + u' + \frac{V'^2}{2}\right) = \frac{d(W\varrho_1 u_1)}{dt} - \phi \tag{8}$$

where Q is the mass flow rate entering the vessel, p', ϱ', V', u' are respectively the entering fluid pressure, density, velocity and internal energy per unit mass; ϱ_1, u_1 are the density and the internal energy per unit mass of the fluid in the vessel, and ϕ is the heat flow per unit time entering the vessel.

— For pneumatic actuators, if one considers W_1 as the increasing chamber volume, and ϕ' the heat flow per unit time entering into it, he has

$$Q'\left(\frac{p'}{\varrho'} + u' + \frac{V'^2}{2}\right) = p_1 A_1 \dot{s} + \frac{d(W_1 \varrho_1 u_1)}{dt} - \phi' \tag{9}$$

where Q' is the entering mass flow rate; p_1, A_1 and $\dot{s}$ are respectively the fluid pressure inside the chamber, the chamber area and the piston speed.

— For pneumatic actuators, if W_2 is the decreasing chamber volume and ϕ'' the entering heat flow rate, one has

$$Q''\left(\frac{p''}{\varrho''} + u'' + \frac{V''^2}{2}\right) = -p_2 A_2 \dot{s} + \frac{d(W_2 \varrho_2 u_2)}{dt} - \phi'' \tag{10}$$

where Q'', p'', ϱ'', V'', u'' are respectively entering mass flow rate, pressure, density, velocity and internal energy per unit mass; p_2 and A_2 are the fluid pressure in the chamber and the area of the chamber.

Eqs. $(8-10)$ describe the relationships between the rate of energy contributed by the entering fluid to the fluid in the considered vessel or chamber, the rate of increase of the internal energy of the fluid contained in the vessel or chamber with time t, and the heat flow rate contributed to the same fluid.

Ducts where non permanent fluid flow occurs are now considered. It is convenient to use the first law of thermodynamics in the form

$$\delta Q_M + \delta E = du + pd\left(\frac{1}{p}\right) \tag{11}$$

where δQ_M is the input energy per unit fluid mass due to heat injection; δE is the input energy per unit fluid mass due to heat produced by friction; u, p, ϱ are respectively the fluid internal energy per unit mass, pressure and density. The symbol δ is used when differential quantities are not exact differentials.

With heat flow per unit mass and unit time designated as Φ, one has

$$\frac{\delta Q_M}{dt} = \Phi. \tag{12}$$

The fluid temperature T is defined through the equation of state

$$T = \frac{p}{R\varrho} \tag{13}$$

where R is the air gas constant, $R = c_p - c_v$, that is the difference between the specific heats at constant pressure and volume. Calling $k = c_p/c_v$ the specific heat ratio, (13) gives

$$T = \frac{p}{\varrho} \frac{k}{c_p(k-1)}. \tag{14}$$

By using Eqs. (7) and (12) one obtains

$$\delta Q_M = -\alpha(T - T_w) \frac{4}{\varrho D} dt. \tag{15}$$

The friction energy δE is generated by transformation of the mechnical work carried out by the fluid, flowing at the mean speed V, against the fluid resistence in the duct. If f is the Darcy friction coefficient, the resistance is equivalent to a force F applied by the duct wall to the moving fluid, that is given, for unit fluid mass and unit duct length, by the expression

$$F = \frac{fV^2}{2D}. \tag{16}$$

The work of friction forces per unit time, is equal to the product of F with the absolute value of the mean speed, $|V|$, i.e.,

$$\frac{\delta E}{dt} = \frac{f |V^3|}{2D}. \tag{17}$$

Developing Eq. (11), with $du = c_v\, dT$, gives

$$\delta Q_M + \delta E = du + pd\left(\frac{1}{\varrho}\right) = c_v\, dT - \frac{p}{\varrho^2}\, d\varrho. \tag{18}$$

Taking into account expression (13), one receives after differentiation

$$dT = \frac{1}{R\varrho}\, dp - \frac{p}{R\varrho^2}\, d\varrho. \tag{19}$$

Introducing expressions (15), (17), (19) in Eq. (18), yields

$$-\alpha(T - T_w) \frac{4}{\varrho D}\, dt + \frac{f |V^3|}{2D}\, dt = \frac{c_v}{R\varrho}\, dp - \frac{c_v p}{R\varrho^2}\, d\varrho - \frac{p}{\varrho^2}\, d\varrho. \tag{20}$$

Taking into account the relations $c^2 = kRT = k(p/\varrho)$, where c is the sound speed in air at temperature T, the second member of Eq. (20) can be expressed as follows:

$$\frac{kp}{\varrho} \frac{1}{kp} \frac{c_v}{(c_p - c_v)}\, dp - \left(\frac{c_v}{c_p - c_v} + 1\right) \frac{kp}{k\varrho} \frac{d\varrho}{\varrho} = c^2 \left(\frac{1}{(k-1) k} \frac{dp}{p} - \frac{1}{k-1} \frac{d\varrho}{\varrho}\right). \tag{21}$$

Introducing Eq. (21) in (20), gives the differential relationship connecting ϱ and p, i.e.,

$$\frac{d\varrho}{\varrho} = -\frac{f |V^3|}{2D} \frac{k-1}{c^2}\, dt + \frac{4\alpha}{\varrho D} \frac{k-1}{c^2} (T - T_w)\, dt + \frac{1}{k} \frac{dp}{p}. \tag{22}$$

4 Characteristic equations

The wave propagation in ducts is studied by means of the so-called "characteristic equations" whereby a hyperbolic problem defined by partial differential equations can be transformed into a problem defined by ordinary differential equations that are valid along some given lines. The characteristic equations are established starting from the problem constitutive equations which are the fluid momentum equation and the fluid continuity equation.

The momentum equation states the equilibrium of forces acting on the fluid contained between two adjoining sections of the duct. If A is the duct inner area, and one considers a unit length between two sections (the unit length may be extremely small), the fluid mass contained between the sections is ϱA. The acceleration of this mass is dV/dt, where the velocity V is a function of both the distance x measured along the duct and of time t. Thus

$$\frac{dV}{dt} = \frac{\partial V}{\partial t} + \frac{dx}{dt}\frac{\partial V}{\partial x} = \frac{\partial V}{\partial t} + V\frac{\partial V}{\partial x}.$$

The difference of pressure forces on the adjoining sections is

$$-\frac{\partial(Ap)}{\partial x} = -p\frac{\partial A}{\partial x} - A\frac{\partial p}{\partial x}.$$

If the section area changes along the duct (e.g. by expansion due to pressure increase), the derivative of the area must be taken into account. On the other side, in this case, the fluid between the sections receives from the duct walls a push force due to wall enlargement, that is

$$+p\frac{\partial A}{\partial x}.$$

The net pressure contribution to the force equilibrium equation is therefore

$$-A\frac{\partial p}{\partial x}.$$

The friction force applied by the duct walls on the fluid mass is given by the Darcy formula, and is equal, per unit mass, to

$$-\frac{fV|V|}{2D}$$

where the absolute value applies because this force is always opposite to the velocity direction. The friction force on the mass is

$$-\frac{fV|V|}{2D}(\varrho A).$$

The momentum equation is therefore

$$-(\varrho A)\left(\frac{\partial V}{\partial t} + V\frac{\partial V}{\partial x}\right) - A\frac{\partial p}{\partial x} - (\varrho A)\frac{fV|V|}{2D} = 0. \tag{23}$$

Equation (23) is written in canonical form as follows:

$$\frac{\partial V}{\partial t} + V \frac{\partial V}{\partial x} + \frac{1}{\varrho} \frac{\partial p}{\partial x} + f |V| \frac{V}{2D} = 0. \tag{24}$$

The fluid continuity equation may be obtained by applying the principle of mass conservation to the same small mass as considered above. The difference between mass inflow and outflow through sections at unit (extremely small) length is

$$\frac{\partial(\varrho VA)}{\partial x}.$$

The time variation of the fluid mass between the sections is

$$\frac{\partial(\varrho A)}{\partial t}.$$

The canonical form of the continuity equation, is thus obtained as follows:

$$\frac{\partial}{\partial x}(\varrho VA) + \frac{\partial}{\partial t}(\varrho A) = 0. \tag{25}$$

By developing the derivatives in Eq. (25), and considering that the section A may change with pressure, one obtains

$$\varrho \frac{V}{A} \frac{\partial A}{\partial p} \frac{\partial p}{\partial x} + \varrho \frac{\partial V}{\partial x} + V \frac{\partial \varrho}{\partial x} + \frac{\varrho}{A} \frac{\partial A}{\partial p} \frac{\partial p}{\partial t} + \frac{\partial \varrho}{\partial t} = 0. \tag{26}$$

The area elasticity modulus of the duct wall is denoted as β, viz.:

$$\beta = A \frac{\partial p}{\partial A}. \tag{27}$$

Introducing this symbol in Eq. (26) gives

$$\frac{1}{\beta} \left(\frac{\partial p}{\partial t} + V \frac{\partial p}{\partial x} \right) + \frac{1}{\varrho} \left(\frac{\partial \varrho}{\partial t} + V \frac{\partial \varrho}{\partial x} \right) + \frac{\partial V}{\partial x} = 0. \tag{28}$$

Now, if one adds and subtracts member to member from Eq. (28) the product of a factor λ by Eq. (24) he obtains

$$\left[\frac{\partial V}{\partial t} + (V \pm \lambda) \frac{\partial V}{\partial x} \right] + \left[\frac{1}{\varrho} \frac{\partial p}{\partial x} \pm \lambda \left(\frac{1}{\beta} + \frac{1}{kp} \right) \left(\frac{\partial p}{\partial t} + V \frac{\partial p}{\partial x} \right) \right] + \frac{fV|V|}{2D}$$

$$\pm \lambda \left[\frac{f|V^3|}{2D} \frac{k-1}{c^2} + \frac{4\alpha}{\varrho D} \frac{k-1}{c^2} (T - T_w) \right] = 0. \tag{29}$$

If one puts

$$\frac{1}{\beta^*} = \frac{1}{\beta} + \frac{1}{kp} \tag{30}$$

$$\lambda = \sqrt{\beta^*/\varrho} \tag{31}$$

$$\frac{dx}{dt} = V \pm \lambda \tag{32}$$

he obtains the relationships that are valid along lines defined in a space-time diagram by Eq. (32). The double signs in Eqs. (29) and (32) have a one-to-one correspondence.

In addition to the relationship between density and pressure, the relationship between density and pressure, which is given by Eq. (22), must also be considered.

The simpler equations for adiabatic ducts were obtained by the authors in Refs. [15 – 16].

5 Program structure and simulation

In general, each pneumatic element used, in realizing a pneumatic circuit, has two types of parameters: 1) typical parameters for the component, which are defined before and independently of future use; 2) parameters which depend on the way these components are used, or which in other words are typical of the circuit created.

5.1 Component definition

The parameters of the components, that are defined by the point 1) above, must thus be entered at component definition level, and those defined by the point 2) must be entered at circuit generation level. For a propagative module, for example, the parameters of the first kind will include diameter, friction coefficient, modulus of compressibility, but not length, wall temperature, initial pressure, number of subdivisions, etc. ..., which are parameters of second kind.

As the program can simulate a circuit with any set of components (after verifying that they are compatible), a distinction must be made between *ordinal number and subtype*. The *ordinal number* labels each component used to create a circuit and increases with circuit development; the *subtype number* is used to define at which component kind the module belongs.

Parameters for components are memorized in matrices or vectors of suitable size, which will in turn form the "component data" file, where each component is identified by its subtype number. This file has the same function which in practice is performed by the component store containing pneumatic parts ready for use in circuit construction.

Friction force in actuators can be simulated with the following heuristic equation presented in Ref. [17] and experimentally confirmed:

$$F_f = F_a + [1 + K_1 \dot{s}^\gamma] [K_2 |p_1 - p_2| + K_3 p_2]$$

where F_a is static friction with nil pressure, K_1, K_2, K_3 and γ are constants, p_1 and p_2 are the chamber pressures and $\dot{s}$ is the piston speed.

5.2 Circuit definition

The structure of a circuit is positively identified by a matrix $N(2, X)$ which thus has a fundamental function for both circuit entry and simulation. X is only limited by the computer memory. The circuit is entered as a sequence of sections belonging both to nonpropagative modules or ducts. The number of columns is equal to the number of circuit sections. Matrix N contains section sequence information together with ordinal numbers assigned to modules as they are used. This matrix can then be divided into submatrices, each of which is associated with a module. In matrix N, each module used occupies as many columns as there are sections involved. As the use of resistance and capacity modules involves entering two sections, four matrix elements (two columns) are available. This is thus a 2×2 submatrix. Joints involve three sections, as they connect three ducts, and will thus occupy three columns in matrix N. Actuators involve 4 sections, 8 matrix elements. This is thus a 2×4 submatrix. For the other modules (ducts, module supply, module exhaust, and capacities), 2×1 submatrices are used.

It is thus apparent that the matrix elements associated with each component in this matrix are not always sufficient for memorizing all parameters relating to a component. Consequently other matrices or vectors have to be introduced; the ordinal number for the component is used to "enter" into these matrices or vectors, while the parameters associated with the component are used to "exit" from them. In matrix N, each kind of component is identified by a number, viz.:

exhaust 1, supply 2, duct 3, resistance 4, actuator 5, joint 6, and capacity 7.

For each module, the characteristic submatrix is indicated together with any matrices connected with it.

5.2.1 Exhaust and supply modules

These modules must be dealt with together. The program checks i × f they must be used as "start" module or as "end" module. A "start" module is set at the beginning of a new branch of the circuit and uses a negative characteristic (C^-). A "end" module closes a circuit branch and uses a positive characteristics (C^+). To recognize the two modules, a different number is put in the first row of the N matrix: the value 2 is for the "start" module, the value 1 for the "end" module.

5.2.2 Propagative module

The propagative module involves a number of sections depending on the number of subdivisions in which the duct is splitted. As the first and last sections of a duct belong to the modules connected to the duct, the propagative module will appear explicitly in matrix N only if its subdivision number is greater than 1. For n subdivisions, there are $n - 1$ columns.

5.2.3 Resistive module

Four matrix elements are available, for this module, in matrix N. With the information stored in it, it is possible to individualize all parameters relative to the resistence and to the ducts connected with it.

5.2.4 Actuator module

Four matrix elements are available, two for input and two for output ports. It is possible to individualize the ordinal number of the actuator, the subtype, the inlet port (front or rear) and the ordinal number of the preceding duct, while all other parameters are memorized in other matrices.

5.2.5 Joint module

The joint module has three outlets and takes six matrix elements in N matrix. The joint is the only component which does not need a subtype number. Consequently, parameters for joints need not be entered when defining components. Four, of these six matrix elements, are defined at the time the user enters a new joint, while the remaining two come into play when the module's third outlet is connected to the circuit.

A matrix JR is then generated in parallel to matrix N with three lines and as many columns as there are joints. Matrix JR is "dynamic" in nature, i.e. the elements memorized in it will vary during circuit creation. This matrix is very important because it consents to establish the correct choice of the characteristics for the third outlet port.

5.2.6 Capacitive module

Four matrix elements are available in matrix N in which the identification number, the ordinal number, the subtype and the ordinal number of the preceding duct are memorized.

5.3 Simulation

Each circuit can be considered to consist of propagative elements connecting non-propagative elements. The propagative elements are the ducts and define the domain governed by hyperbolic partial differential equations solved using the characteristics method. The non-propagative modules with the associated equations used to model their behaviour define the boundary conditions for the propagative modules. Once state variables at time i are known, the program calculates state variables at time $i + 1$ by "reading" matrix N exactly as it was generated. It is thus possible to determine the sequence of modules section by section with the associated subtype and ordinal numbers, as well as to "enter" into previous matrices and associate each module with its characteristic parameters. Once velocity at time $i + 1$ for a given section is known, it is necessary to check that it does not exceed local sonic conditions.

It should be noted that sonic conditions in a given section can be reached only in transients. In steady-state conditions, a further expansion of the fluid in the duct is not possible without an increase in fluid velocity (continuity would no longer exist). Sonic conditions may occur systematically only in the terminal section of a duct connected to an exhaust module. In this case the downstream conditions (exhaust environment) are no longer felt in the duct.

In non-sonic conditions, the equations linking the exhaust module to the duct are as follows:

(a) Characteristic equations.
(b) External constant pressure (defined by user).

In non-sonic conditions, it is sufficient to substitute the pressure imposed by outside conditions (b) in the characteristic equations to determine outlet section velocity at time $i + 1$. When the duct reaches sonic conditions in the outlet section, the constant pressure conditions no longer occurs. It is thus necessary to introduce, instead of (b), the condition of sonic speed $V_{i+1,j}^2 = [k(p_{i+1,j})/(\varrho_{i+1,j})]$.

5.4 Management of results and print options

Management of results and associated printout are handled separately in the GEST subroutine, which reads data files generated by the subroutine which performans circuit simulation. These state variables are memorized in a three dimensions matrix $S(I, J, NTG)$, where indices J and NTG represent the section and the time of the associated data respectively. The state variables are pressure, velocity, flows, Mach number, density and temperature.

It is possible to create appropriate graphs for each kind of module. It is possible to follow the development of the state variables in the time for not propagative modules, as well as is possible to freeze time and to analize the state variables in the circuit.

The matrices generated in the data files by circuit simulation are of considerable size. Consequently, it is necessary to save reduced files containing only the informations which the user consider to be useful. It is possible to save reduced files where limitations are placed on minimum and maximum simulation time and on the step with which data are saved.

6 Examples

A system made by a short tube (0.2 m length, 13 mm inner diameter), a valve (1/2″ size, CETOP coefficients $C = 1.86 \cdot 10^{-7} \, m^3/sPa$ (ANR), where ANR is the standard reference atmosphere, according to ISO 8778, and $b = 0.245$), and a long tube (5 m length, 13 mm inner diameter) discharging to an ambient at a given pressure is considered. Both tubes are considered very stiff ($\beta = \infty$). Friction coefficient in tubes is $f = 0.05$. Wall temperature is $T_w = 300$ K. The supply pressure is constant end equal to 9 bar and the supply air temperature is 300 K. The discharge pressure is 7 br and the temperature also 300 K. A fast valve opening was simulated. The diagrams of flow rate, density, pressure, speed, temperature, Mach number along the duct are shown respectively in Figs. 1, 2 and 3 after 50 ms from valve opening.

Another example is given by the charging and discharging of a pneumatic capacity. Here the capacity is given by a volume of 2 litres and the inlet and outlet losses in the capacity are considered given by the CETOP coefficients $C = 1.86 \cdot 10^{-6} \, m^3/sPa$ (ANR) and $b = 0.245$ (as shown by the authors in [16]). Charging or discharging of the capacity occurs through a short duct (0.3 m length, 13 mm inner diameter, $f = 0.05$, $\beta = \infty$), a valve (that is described by the same CETOP parameters like the valve described in the first example), and another short duct like the first one but of 0.1 m length. In case of charging, supply pressure is 9 bar and initial internal pressure in the volume is ambient pressure. In case of discharging, inner pressure in the vessel is 9 bar, and discharge pressure is ambient pressure. Temperature of supply air and walls is 300 K.

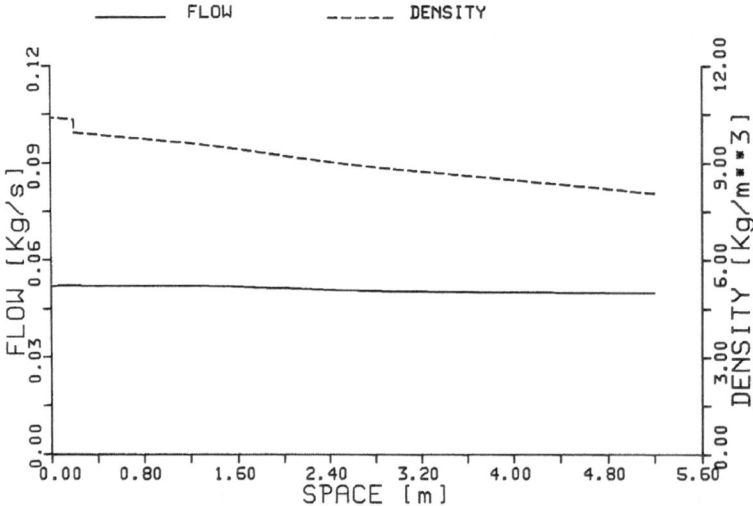

Fig. 1. Variation of flow throughput and density in space after 50 ms

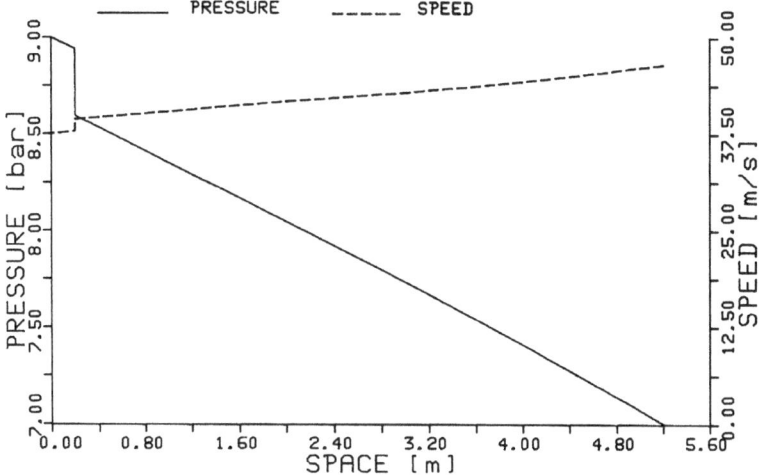

Fig. 2. Variation of pressure and speed in space after 50 ms

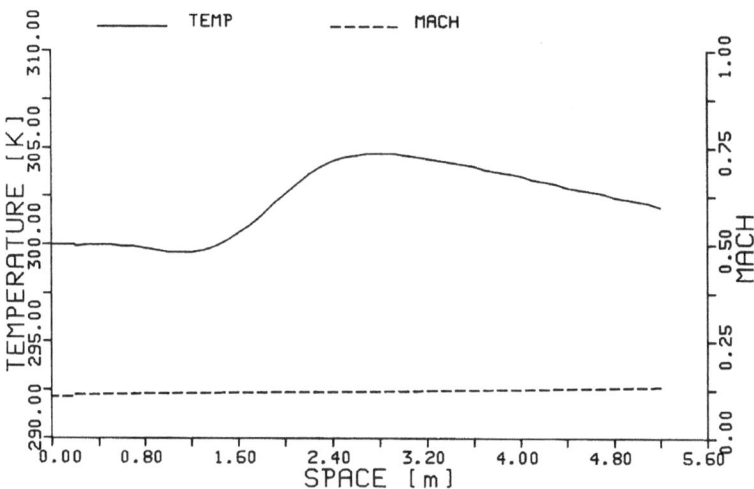

Fig. 3. Variation of temperature and Mach number in space after 50 ms

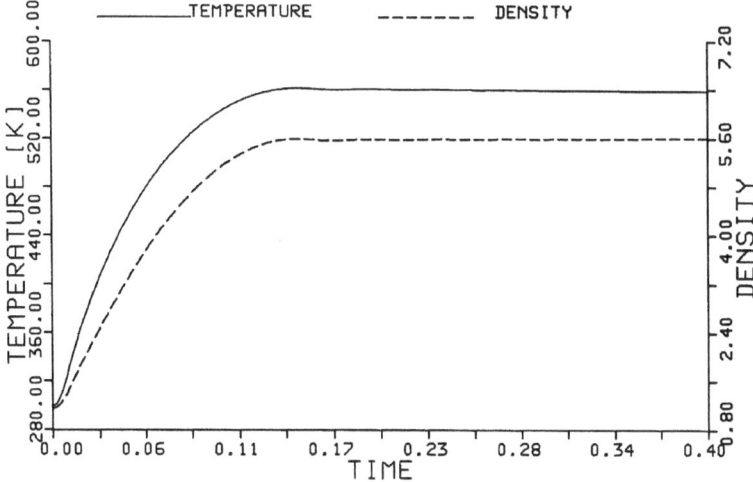

Fig. 4. Temperature and density values during volume charging

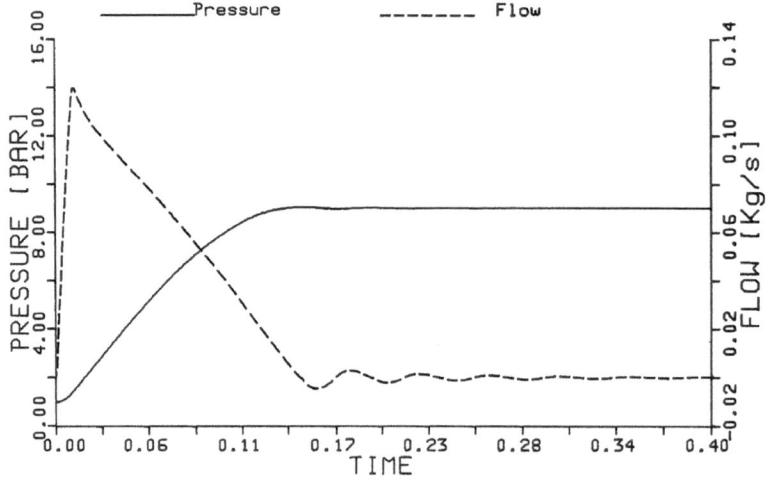

Fig. 5. Pressure and inflow values during volume charging

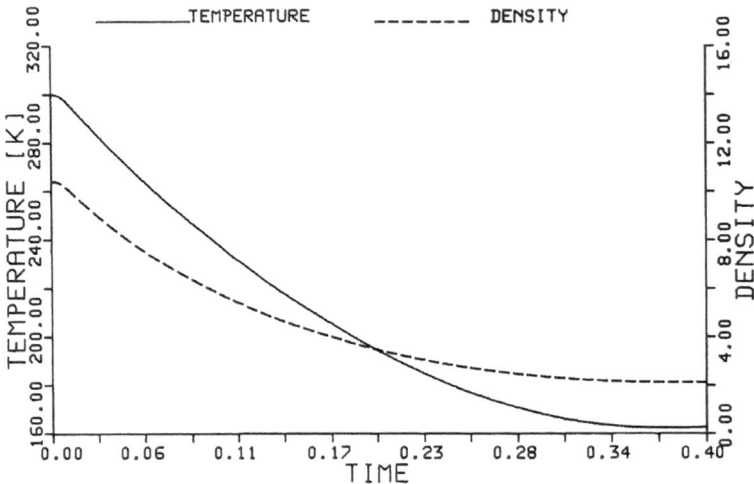

Fig. 6. Temperature and density values during volume discharging

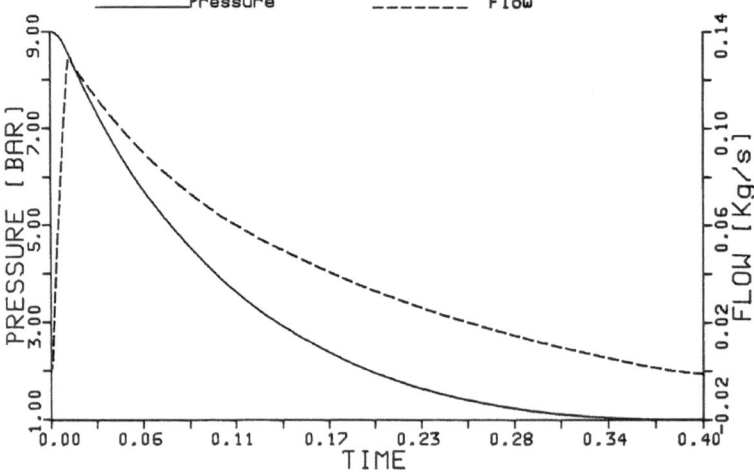

Fig. 7. Pressure and outflow values during volume discharging

Figures 4 and 5 show the time variation of temperature, density, pressure and inflow in the vessel in case of charging, while Figs. 6 and 7 show the same quantities for discharging, but the flow is obviously reversed. The considered volume is cylindrical with length double than the diameter. The heat transfer coefficient in this case is considered constant and equal to $\alpha = 10 \text{ W/m}^2\text{K}$.

7 Conclusions

In addition to examples discussed above, a number of different pneumatic systems were examined with different initial conditions, and the computerized results were compared with experimental results. Apart from the first microseconds from the beginning of a transient phenomenon, when shock wave effects are strong, the two sets of results showed good agreement, even if the systems were quite complicated. This fact demonstrated that the method presented here is capable of providing an accurate solution to any problems involving pneumatic circuit transients, with very good time definition.

For example, braking time delay in trailing trucks due to transmission of braking power through a long pneumatic line and connected mechanical devices has been determined with a very good degree of precision. In addition, the behaviour of high speed vehicle active pneumatic suspensions with respect to time has been determined for various vehicle speed and track curvature and irregularities (the electronic control unit behaviour was also taken into account).

The method appears to be suitable for a great number of other cases involving circuits and systems through which a fluid is distributed, or where a power operates or a control fluid is operated. The method can accurately define the time behaviour of any system component during transient phenomena.

Acknowledgments

This study was partly sponsored by the Italian Ministry of University.

References

[1] Backè, W., Hoffman, W.: DSH-program system for digital simulation of hydraulic systems. In: Proceedings of 6th Int. Fluid Power Symposium, pp. 95–114, Cambridge, UK 1981.

[2] Krus, P., Palmberg, J. O.: Simulation of fluid power systems in time and frequency domains. In: Proceedings of 7th Int. Fluid Power Symposium, pp. 73–79, Bath, UK 1986.

[3] Vilenius, M. J., Luomaranta, M. K., Rinkinen, J. A.: CATSIM — a new type of CAD program for hydraulic circuit design. In: Proceedings of 7th Int. Fluid Power Symposium, pp. 65–72, Bath, UK 1986.

[4] Zhang, H.: HYCAD — a general hydraulic simulation package. The Journal of Fluid Control **8/2**, 36–44 (1988).

[5] Richard, C. W., Tilley, D. G., Tomlinson, S. P., Burrows, G. R.: A second generation package for fluid power systems. In: Proceedings of 9th Int. Fluid Power Symposium, pp. 315–322, Cambridge, UK 1990.

[6] Watton, J., Salters, D. G., Nelson, R. J.: New software techniques for the CAD of fluid power systems. In: Proceedings of 9th Int. Fluid Power Symposium, pp. 323–330, Cambridge, UK 1990.

 [7] Handroos, H. M.: A library of component models in a fluid power circuit performance simulation program-aspects and methods for improving. In: Proceedings of 9th Int. Fluid Power Symposium, pp. 331–342, Cambridge, UK 1990.
 [8] Manning, J. R.: Computerized method of characteristics calculations for unsteady pneumatic line flows. J. Basic Eng. **90/3**, 231–240 (1968).
 [9] Tsao, S.: Numerical solution of transients in pneumatic networks — transmission-line calculations. J. Appl. Mech. **35/3**, 588–595 (1968).
[10] Tsao, S.: Numerical solution of transients in pneumatic networks — Part 2 — non linear termination problems. J. Appl. Mech. **36/3**, 588–593 (1969).
[11] Tsao, S., Rodgers, W.: Numerical solution of transients in pneumatic networks — Part 3 — network problems with branching. J. Appl. Mech. **36/3**, 594–597 (1969).
[12] Yoshioka, M.: Infinite series representation of transient response in fluid transmission lines. The Journal of Fluid Control **15/4**, 18–41 (1984).
[13] Backè, W., Ohligschläger, O.: A model of heat transfer in pneumatic chambers. The Journal of Fluid Control **20/1**, 61–78 (1989).
[14] Hausen, H.: Z. Ver.deut. Ingr., Beih. Verfahrenstech. **4**, 91–98, 1943. In: Eckert, E. R. G., Drake, R. M.: Heat & Mass Transfer, 2nd ed., p. 212. New York–Toronto–London: McGraw-Hill 1959.
[15] Romiti, A., Raparelli, T.: Rigorous analysis of transients in gas and liquid circuits and comparison with experimental data. Torino: Mechanical Department 1990.
[16] Romiti, A., Raparelli, T.: Dynamic modelling and simulation of pneumatic systems. In: Fluid Power (Fairhurst, R. M., ed.), pp. 137–146, 9th Int. Symposium on Fluid Power, Cambridge, UK 1990.
[17] Belforte, G., D'Alfio, N., Raparelli, T.: Experimental analysis of friction forces in pneumatic cylinders. The Journal of Fluid Control **20/1**, 42–60 (1989).

Authors' address: Prof. A. Romiti, Dipartimento di Meccanica, Politecnico di Torino, C.so Duca Degli Abruzzi 24, I-10129 Torino, Italy.

Acta Mechanica (1992) [Suppl] 3: 99–114

Approximate dynamic responses in random media

G. Dasgupta, New York, U.S.A.

Summary. Stochastic constitutive properties are important in *real-world* dynamic analyses. The unconventional field equations, with spatially varying random coefficients, pose a computational challenge. For mild stochasticity, straightforward perturbations yield satisfactory response statistics. On the other hand, large statistical deviations can be captured from infinite sequence approximations, with a stationary iteration scheme introduced by Boley. For coupled thermal and elastodynamic random field problems, only the "inversions" of the uncoupled operators pertaining to a uniform continuum are deemed adequate. Stochastic boundary element approximations follow directly from the proposed stochastic Green's functions. In the light of Tatarski's wave considerations, corresponding stochastic finite element formulations require additional convection like spatially random terms related to *stochastic shape functions*. Crucial aspects of dynamic computations with random material properties are summarized here with sample numerical examples.

1 Introduction

Engineering systems are designed with a variety of materials in a multitude of shapes. During its service life, a system encounters a wide spectrum of time dependent mechanical and thermal loads. The overall performance rating of a system is of major computational interest. In design optimization, the quantitative as well as qualitative questions for structural safety pertain to survival, failure to comply with specifications and undesirable performance such as excessive vibration and thermal degradation. The chances that certain criterion will not be met during the life of the system are of concern. The entire design procedure is thus rendered to be a decision making activity with statistical considerations.

During the numerical computation of dynamic responses, first the constitutive random behavior for the solid or fluid medium is identified. Secondly, the extent of assumptions within linear or nonlinear regime is examined. This is highlighted by the fact that a nonlinear dynamic opertor demands a stochastic description even when deterministic systems are excited by random forces. The subsequent dynamic analysis is required to be performed adhering, as closely as possible, to exact solutions. However, there is no standard algorithmic procedure in the classical *strong* method to solve initial-boundary value problems. As in the case of obtaining particular integrals, conventional direct solution schemes demand different expressions — based on experiences — for forcing functions with different forms of spatial dependency. Likewise, the hyperbolic equations of elastodynamics would have to be treated differently from the parabolic equation of heat conduction. Consequently, the coupled thermo-elastodynamic responses could not be recovered from the existing knowledge of the separate solution procedures for dynamic and temperature distribution problems.

In order to circumvent the aforementioned difficulties with a problem specific technique, Boley in early fifties, [1], explored algorithms which should remain independent of the nature of the governing linear operator, to be indicated by $\mathscr{L}$. A (non-classical) *weak form*, as an integral representation, was advocated to preserve the continuity of solutions within the entire domain Ω. The Green's function $\mathscr{G}$ provided the kernel. Let $\mathscr{L}(x, t)$ be the dynamic linear operator, where x denotes the spatial coordinates and t indicates time, govern the response $u(x, t)$. In case of a forcing function $f(x, t)$ exciting the system, the solution $u(x, t)$ was sought in the following form:

$$u(x, t) = \int_0^t \left[\int_\Omega f(\eta, \tau) \, \mathscr{G}(x, t \,|\, \eta, \tau) \, d\eta \right] d\tau \tag{1}$$

where

$$\mathscr{L}(x, t) \, \mathscr{G}(x, t \,|\, \eta, \tau) = \delta(x - \eta) \, \delta(t - \tau) \tag{2}$$

and δ denotes the Dirac delta. In the notion of the convolution algebra, $\mathscr{G}(x, t \,|\, \eta, \tau)$ is called the *weak* inverse of the operator $\mathscr{L}$ since δ plays the role of the identity. The above temporal integration can be carried out in the Duhamel form. Thust the main task is to quantify the spatial distribution of the aforementioned Green's function $\mathscr{G}$. For some extremely limited cases, e.g. Laplace's potential (when $\mathscr{L} = \nabla^2$), the closed-form expressions for the Green's functions $\mathscr{G}$ were available. To extend the applicability of the Green's function technique to the entire class of time dependent problems, Boley in [1] employed the Liouville-Neumann operator expansion — the infinite series method. Along side he introduced the construction of an *infinite sequence* as well. At **every** step, the zero-th order trial solution $\mathscr{G}^o$ — i.e. the "first guess" (as termed in [1], p. 251 for Eq. 8), was utilized to be the kernel in the integral representation. Essentially, the required Green's functions $\mathscr{G}$ was represented as follows:

$$\mathscr{G} = \mathscr{G}^o + \sum_{i=1}^\infty \mathscr{G}^{(i)} \quad \text{where} \quad \mathscr{G}^{(i)} = \mathscr{P}[\mathscr{L}, \mathscr{G}^o] \, \mathscr{G}^{(i-1)} \quad \text{with} \quad \mathscr{G}^{(0)} = \mathscr{G}^o \tag{3}$$

Boley's papers [1] and [2] provide and illustrate the construction of the operator $\mathscr{P}$ in Eq. (3). It is stationary since it does **not** change with the index i. Boley also discussed in [1] the convergence criterion for the stationary iteration implemented in the aforementioned infinite sequence method. The extraordinary breadth of applications of Boley's technique should be noted. Starting with a *partically complete* first guess $\mathscr{G}^o$, it is possible, even in one-dimensional problems, to recover the complete solution $\mathscr{G}$ comprising of both the incoming and outgoing fluxes with the latter as the first guess. This observation is particularly important in dynamic problems which involve both finite and unbounded bodies. Soil-structure interaction problems [3] can be cited to be an important area of such applications.

In the case of three-dimensional elasticity the Green's function construction is rather involved due to the vectorial nature of the displacement potentials. Boley in [2], demonstrated the application of the methodology sketched in [1] for the homogeneous halfspace problem with a buried force. For the randomly distributed material properties such fundamental solutions have been used in [4] by the author as the *first guess* to get the appropriate Green's functions for more complicated problems at hand. Furthermore, using the perturbation scheme [5] with computer algebra, the author implemented the Liouville-Neumann expansion, described in [1], around deterministic operators in [6].

For dynamic problems, with randomly varying coefficients originating out of statistically simulated samples, this strategy of *stationary iteration* with deterministic $\mathscr{G}^o$ for homogeneous fields is applied numerically by the author in this paper. Furthermore, it has been shown that there is no restriction of the methods proposed by Boley in [1] for differential equations with stochastically *large* parameters.

The Liouville-Neumann expansion in [1] yielded repeated integration over the domain of interest. The equivalent differential formulation was considered by Tatarski, in [7], for random media. In the force-deformation relationship, the primary expressions, associated with the *first guess*, as in [1], pertain to the homogeneous system. The effect of large stochastic variability is witnessed in the terms which contain the gradient of the stochastic field. These additional convective expressions are indeed closely related to the integral refinements in the infinite sequence constructed by Boley in [1] and [2].

Even common engineering objects are so complex in geometry and/or in constitutive inhomogeneity that the aforementioned algebraic schemes become intractable. These complexities preclude the direct application of the continuum methods of Boley [1] and [2] and of Tatarski [7]. Thus in practice, based on the knowledge of exact solutions pertaining to systems of rather simplified geometry and uniform material property, alternative discrete approximations are pursued. Therein the integration of partial differential equations of mathematical physics, which dictate motion and temperature variation, is reduced to the solution of rather simplified algebraic equations. Consequently, the response evaluation proceeds in numerical steps amenable to digital computer programming. With the advent of high speed computers, finite difference, finite element and boundary element techniques thus emerged as approximate but very powerful calculation tools. There has been an upsurge in interest in formulating numerical solution schemes to solve thermo-mechanical systems where the physical variables can, at best, be known only in the statistical sense. Eventually, it became possible to carry out, rather inexpensively, very large scale thermo-mechanical simulations for realistic engineering systems.

Historically, rigorous probabilistic computation and the associated mathematically sound research work pertained to random fluid media. For example, a NASA publication [8] of 1968 contains a number of papers of fundamental interest. Atmospheric research contributed substantially in developing dynamic numerical techniques. During wave propagation through turbulent fluids, modeled as stochastic media, the effects of *multiple scattering* received special attention. The propagation of the shock-front indicated substantial thickening which can only be estimated via statistical means. Subsequently, the population mean, statistical dispersion as well as the higher order *statistical moments* of the scatter were estimated. The outstanding feature of these stochastic formulations was that in addition to the conventional deterministic terms, *convective* type randomly fluctuating terms, derived in [7], dominated the calculations.

Efforts were also focused to incorporate realistic random parameters into the structural mechanics problems. In their pioneering work to utilize the finite element method in stochastic dynamic computations, Astill, Nosseir and Shinozuka [9] implemented the Monte Carlo technique to study failure of concrete cylinders under impact loading. Very encouraging results for transient responses, with a population size of only one hundred, were reported. Since each sample was separately discretized and independently computed, the overall statistical results were acceptable. However, while the accuracy was satisfactory, the numerical efficiency was questionable.

In the case of mild stochasticity, i.e. when the gradient of the randomly fluctuating field is rather small, the perturbation formulation, as elaborated in [5], yielded considerably efficient algorithms. Within the finite element setting, a systematic development containing adequate mathematical details was furnished by Nakagiri and Hisada since 1979, in research notes [10] and in a text book [11]. This technique was demonstrated to demand the factorization of the reference "dynamic stiffnes matrix" in a time marching explicit or an implicit scheme only once in order to evaluate any order of statistical moment. Evidently a qualitative improvement over the sample by sample Monte Carlo technique was achieved.

Nakagiri and Hisada, and in the straightforward Monte Carlo simulations, Astill, Nosseir and Shinozuka confined their attention to deterministic finite element shape functions. The dynamic force-deformation relation with *rapidly varying* random material properties without exception demands *stochastic shape functions*, [12]. The correlation between the gradient of the material constitutive field and the strain profile can only be captured via the random convective like terms for the subsequent reliability calculations. Hence, the introduction of stochastic shape functions for finite elements [12] and stochastic Green's functions for boundary elements [4] becomes the most outstanding feature in the field of stochastic computational mechanics. If the samples are ordered "according to the energy norm" the accompanying Monte Carlo simulation technique can be substantially accelerated, as elaborated in [13].

A nondeterministic system subjected to a set of deterministic loads, i.e. *system stochasticity* problems with spatially random material inhomogeneity, is of interest in this paper. Of course, the most involved problem is to solve a stochastic system subjected to a nondeterministic force history. Computationally, this combined effect poses technical rather than conceptual difficulties over system stochasticity problems. Hence, the stochasticity in material properties, which modifies the dynamic "force-displacement" relationship, is of principal interest here. The mathematical bases, specifically suited for the application in the stochastic finite element and stochastic boundary element methods, can be found in the text book by Sobczyk [5] and in a proceeding [14] edited by Keller and McKean. Therein statistical quantities related to the randomness for velocities, strains, displacements, and all relevant responses were elaborated. Lax in [14] described a mathematical treatment of wave propagation and conduction problems in random media. Besides analytical considerations the Monte Carlo simulation technique, method of statistical moments and smoothing schemes were also presented. Based on these theoretical considerations, the crucial mathematical steps for stochastic boundary element and stochastic finite element methods will now be furnished.

2 Symbols and notations

The following terms: *random, nondeterministic, statistical, probabilistic* and *stochastic*; will be used here, somewhat loosely, in a synonymous fashion. In general, for any symbol Z its stochastic counterpart will be designated with a superscript tilde, i.e. by $\tilde{Z}$. The mean (or the expected value) of $\tilde{Z}$ will be denoted by $\bar{Z}$ or in terms of the *expected value operator*, $\mathscr{E}$ as

$$\bar{Z} = \mathscr{E}[\tilde{Z}]. \tag{4}$$

A reference value of $\tilde{Z}$ will be indicated by Z^o, then Z^* will denote the stochastic fluctuating part obeying:

$$\tilde{Z} = Z^o + Z^*. \tag{5}$$

Boley's infinite sequence method [1] and [2] is applied to recover Z^* by iteration using Z^o. In the case when $[\tilde{Z}]$ is a stochastic operator its equivalent counterpart for the mean solution field will be indicated by $\hat{Z}$. Thus for a field variable $\tilde{z}$:

$$\mathscr{E}[[\tilde{Z}]\,\tilde{z}] = [\hat{Z}]\,\bar{z}. \tag{6}$$

Except for trivially small stochasticity:

$$[\hat{Z}] \neq [\bar{Z}]. \tag{7}$$

For example, the deterministic stress (σ)-strain (ε) relationship

$$\varepsilon = \frac{\sigma}{E} \tag{8}$$

involving the modulus of elasticity E cannot be "extended", in the *ad hoc* fashion for the mean values, i.e.

$$\bar{\varepsilon} \neq \frac{\bar{\sigma}}{\bar{E}}. \tag{9}$$

The pioneering work of Keller [15] and [16] furnished a convenient estimation of the mean stochastic operator $[\hat{Z}]$. The solution to this operator indicates an immediate stochastic boundary element formulation. In the corresponding finite element formulation, the contribution of the randomly fluctuating expressions captured via the *stochastic shape functions* yield additional *stochastic convective* matrices, similar to those terms mentioned in [5] and [7], which indeed validate Eq. (7).

3 Stochastic boundary element formulation

The boundary element system matrices can be derived from the associated Green's functions. The general procedure is detailed in the text book [17], in the computer program BEASY [18] and in research publications [19] and [20]. In the class of familiar *random vibration* problems the Green's functions are deterministic. The system stochasticity problem renders fluctuations in the operator itself. The equivalent operators $[\mathscr{L}]$ for smoothed responses are of interest here for practical computations.

The thermo-mechanical field equations, stated in Sobczyk [5] p. 113, are rewritten as follows. The stochastic constitute parameters are:

$\tilde{\alpha}$: coefficient of linear expansion; $\tilde{\lambda}$ and $\tilde{\mu}$: the Lamé constants; $\tilde{\varrho}$: density of the medium; $\tilde{s}$: specific heat, and $\tilde{h}$: coefficient of thermal diffusion. Deterministic constitutive and field variables, Z in [5] p. 113 are replaced by their stochastic counterparts, $\tilde{Z}$, yielding:

$$\tilde{\varrho}\ddot{\tilde{u}} = \tilde{\mu}\nabla^2\tilde{u} + (\tilde{\lambda} + \tilde{\mu})\,\nabla(\nabla \cdot \tilde{u}) + \nabla(\tilde{\lambda}\nabla \cdot \tilde{u}) + (\nabla\tilde{\mu}) \times (\nabla \times \tilde{u}) + 2(\nabla\tilde{\mu} \cdot \nabla)\,\tilde{u} - \nabla((\tilde{\lambda} + \tilde{\mu})\,\tilde{\alpha}\tilde{T}) + \tilde{f} \tag{10}$$

and

$$\tilde{\varrho}\tilde{s}\dot{\tilde{T}} = \nabla \cdot (\tilde{h}\nabla\tilde{T}) - (\tilde{\lambda} + \tilde{\mu})\,\tilde{\alpha}T_o\nabla\dot{\tilde{u}} + \tilde{q}. \tag{11}$$

Therein the gradient operator is ∇ in the spatial coordinates $\{x\}$; and a superscript dot indicates once differentiation with respect to time t. The stochastic body force is $\tilde{f}$, the stochastic heat source is $\tilde{q}$, and T_o denotes the prescribed initial temperature distribution. In this paper on system stochasticity the "forcing functions" will be illustrated with a deterministic time harmonic vector $\{g(t)\}$:

$$\{g(t)\} = \begin{Bmatrix} f_a \\ q_a \end{Bmatrix} \exp(-i\omega t) \quad \text{where} \quad \{g_a\} = \begin{Bmatrix} f_a \\ q_a \end{Bmatrix}. \tag{12}$$

The stochastic amplitude of a steady wave with a frequency ω is denoted by $\tilde{z}_a$. The response vector:

$$\{\tilde{v}_a\} = \begin{Bmatrix} \tilde{u}_a \\ \tilde{T}_a \end{Bmatrix} \tag{13}$$

houses the displacement and the temperature variables, and is governed by:

$$[\tilde{L}]\{\tilde{v}\} = \{g\} \quad \text{where} \quad \tilde{L} = \begin{pmatrix} \tilde{L}_{11} & \tilde{L}_{12} \\ \tilde{L}_{21} & \tilde{L}_{22} \end{pmatrix}. \tag{14}$$

The nondeterministic field operator is decomposed such that:

$$\tilde{L} = L^0 + L^*, \quad \text{where} \quad L^0 = \begin{pmatrix} L_{11}^o & 0 \\ 0 & L_{22}^o \end{pmatrix}, \quad \text{and} \quad L^* = \begin{pmatrix} L_{11}^* & \tilde{L}_{12} \\ \tilde{L}_{21} & L_{22}^* \end{pmatrix}. \tag{15}$$

The operators with the superscript "o" pertain to the *uncoupled thermo-mechanical field equations* with constant constitutive coefficients. Such a splitting of the stochastic operator naturally adapts to Boley's stationary iteration computations [1] in a stochastic boundary element formulation. The useful Green's function relationship for the first guess $[G^o]$ as in [1] is thus given by:

$$[L^0]^{-1} = [G^0] = \begin{pmatrix} G_{11}^o & 0 \\ 0 & G_{22}^o \end{pmatrix}. \tag{16}$$

An existing boundary element code, e.g. BEASY [18], which solves the steady vibration problems, and the temperature problems separately will have the modules to generate and use the boundary element system matrices related to G_{11}^o and G_{22}^o, respectively.

In order to evaluate the dynamic response statistics by a Monte Carlo simulation, it is numerically more efficient to generate the reference dynamic system (with superscript "o") for the largest values of the stiffness type constitutive properties. Then, the energy ordering scheme of [13] is extremely fast to solve for the sample space with the stationary iteration scheme. Unlike the perturbation method, the proposed strategy is indeed convergent for large stochastic deviation of material properties.

The aforementioned formulation becomes simple for the case of mild stochasticity with the additional assumption of uncoupling between the temperature and the kinetic effects. Perturbation expansion of the operator equation is then possible. The second order approximation, perturbed about the mean, is implemented in the form:

$$[\tilde{L}] \cong L^o + \varepsilon L^{(1)} + \varepsilon^2 L^{(2)} \quad \text{with} \quad \mathscr{E}[L^{(1)}] = 0. \tag{17}$$

Then Keller's formula, [15] and [16], with the *equivalent operator* $[\hat{L}]$ for the meanfield:

$$\{\bar{v}_a\} = \begin{Bmatrix} \bar{u}_a \\ \bar{T}_a \end{Bmatrix} \tag{18}$$

can be directly obtained from:

$$[\hat{L}]\{\bar{v}_a\} = \{g_a\}, \Rightarrow \left[L^o + \varepsilon^2 [\mathscr{E}[L^{(2)}] - \mathscr{E}[L^{(1)}G^oL^{(1)}]] \right] \{\bar{v}_a\} = \{g_a\}. \tag{19}$$

Further simplification, i.e. Bourret approximation for Dyson's equation, is possible if the second perturbed term $L^{(2)}$ in Eq. (17) is negligibly small.

4 Stochastic finite element formulation

A finite element solution strategy, for the randomness in spatial variability of constitutive properties, will now be considered. The aforementioned continuum statements, Eqs. (10) and (11), will be the starting point. The stochastic coupling, mentioned in [7], in the form of the covective terms, i.e. products of the gradient of the random field and the strain type quantities, does not involve the random density of the medium, $\tilde{\varrho}$. Thus it is adequate to sketch out the stochastic finite element method for elastostatic cases. The *stochastic shape functions* could be used to compute the associated consistent mass matrices in a dynamic computation.

The nodal force-displacement relation is expressed via the matrix $[\mathcal{K}]$. The forcing function and the corresponding response are vectors $\{\mathcal{F}\}$ and $\{\mathcal{U}\}$, respectively. Then the familiar deterministic equation $[\mathcal{K}]\{\mathcal{U}\} = \{\mathcal{F}\}$ can be *conceptually* extended to the following general random form:

$$[\tilde{\mathcal{K}}]\{\tilde{\mathcal{U}}\} = \{\tilde{\mathcal{F}}\}. \tag{20}$$

In order to emphasize the construction of the finite element stiffness matrices for system stochasticity the applied force will be kept deterministic:

$$\{\tilde{\mathcal{F}}\} = \{\mathcal{F}\}. \tag{21}$$

The generic equation of interest is then:

$$[\tilde{\mathcal{K}}]\{\tilde{\mathcal{U}}\} = \{\mathcal{F}\}. \tag{22}$$

The stiffness of the system is modeled as a random quantity, i.e. $\mathcal{K} \to \tilde{\mathcal{K}}$, which entails the associated response to be $\tilde{\mathcal{U}} \leftarrow \mathcal{U}$. This paper focuses on the analysis of uncertainties in responses originating from the assignment of probabilistic description for the material properties.

Currently, a number of related finite element publications report the deployment of random variables with deterministic finite element shape functions. The essential requirement to re-derive the stiffness matrices augmented by convected like terms of [7], which arise out of spatial variability, is essentially ignored when computer codes with the usual deterministic finite element Subroutines are used in material stochasticity calculations. However, for a class of problems, when the system stochasticity reduces to a random loading case, such *short-cuts* incur no error. In such a case of the *stochastic shear beam* the ordinary differential equation governing the random shear angle $\tilde{r}$ is:

$$\tilde{A}(x)\,\tilde{G}(x)\,\frac{d\tilde{r}}{dx} = T(x) \tag{23}$$

or

$$\tilde{b}(x)\,\frac{d\tilde{r}}{dx} = T(x). \tag{24}$$

The transverse load is $T(x)$, and $\tilde{A}(x)$ and $\tilde{G}(x)$ are the random cross sectional area and the random shear modulus, respectively, leading to $\tilde{b}$ to be the random process depicting the stochastic shear stiffness. Vanmarcke and Grigoriu described elegant procedures on the basis of the *scale of fluctuations* leading to the successful finite element formulation of the stochastic shear beam in [21]. Methods of selection for the computer input for the random variable $\tilde{b}$ for an element (i) in a Monte Carlo simulation, are described by Yamazaki and Shinozuka in [22]. It should be emphasized that for a given correlated random process with bounded $\tilde{b}(x)$, the principal component analysis with independent Gaussian random numbers is erroneous.

The system stochasticity problem in Eq. (24) is reducible to a case of random forcing function, i.e.:

$$\tilde{b}(x)\frac{d\tilde{r}}{dx} = T(x) \tag{25}$$

leads to

$$b^o\frac{d\tilde{r}}{dx} = \tilde{T}(x) \tag{26}$$

where

$$\frac{\tilde{b}}{b^o} = \frac{\tilde{T}(x)}{T(x)} \tag{27}$$

and b^o is the constant shear stiffness within the element, which is, customarily, the spatial average. The corresponding finite element calculations can be carried out with linearly varying shape functions. The element formulation is indeed exact. Mild or rapidly varying stochastic material properties can be treated identically. For the i-th element the average value b^o is to be input in a computer code equipped to deal with random loading problems. Similar treatment for the the general self-adjoint problems, e.g. for a truss member, is not possible as described below.

The linear shape functions, are also exact for a one-dimensional truss member where the axial stiffness AE is constant. Therein A and E are the area of cross section and modulus of elasticity, respectively. The area of the bar is assumed to be a constant, A^o. The random modulus of elasticity will be denoted by $\tilde{E}(x)$. The spatial variability in terms of the non-dimensional "unit" random process $\tilde{g}(x)$ can now be defined according to

$$\tilde{g}(x) = \frac{\tilde{E}(x)}{E^o} \tag{28}$$

where E^o is the spatial mean of the process $\tilde{E}(x)$ within the finite element in question. Finally, the displacement equation of equilibrium is obtained by introducing the strain-displacement relation as follows:

$$\widetilde{\frac{d^2\tilde{u}}{dx^2}} + \frac{\overparen{\frac{d\tilde{u}}{dx}\frac{d\tilde{g}}{dx}}}{\tilde{g}} = -\frac{f(x)}{\tilde{g}A^oE^o} \tag{29}$$

where $f(x)$ is the externally applied axial load intensity. The first term is identifed with the usual deterministic formulation. The second one introduces aditional *convected expressions*, due to the spatial variability of the constitutive property. Therein the involvement of the first order derivative of the field response justified the terminology.

The kernel of the system stochasticity is constituted by the random process $\tilde{g}(x)$ of Eq. (28). Hence any term related to $\tilde{g}(x)$ or its derivative $\frac{d}{dx}[\tilde{g}(x)]$ in Eq. (29) cannot be ignored in the associated stochastic finite element formulation. In deriving the matrix equation for the stochastic finite element representation of the truss member Eq. (29) should be considered. This term demands that the stiffness Subroutine in deterministic computer code must be augmented to account for this additional convective term.

An estimation of the stochastic strain-displacement transformation matrix $[\tilde{B}]$ via *stochastic shape functions* can indeed be carried out. Such elaborate and explicit calculations can be economized in a finite element design analysis which relies on a fixed-mesh spatial discretization. Essentially a practical computational strategy is nessecary only to evaluate the strain-displacement transformation matrix. In a conventional deterministic analysis, the deterministic $[B]$ matrix relates the estimated strain field $\{\varepsilon\}$ and the nodal displacement list $\{u\}$ at the element level in the following form:

$$\{\varepsilon\} = [B] \{u\}. \tag{30}$$

The stochastic counterpart will simply be:

$$\{\tilde{\varepsilon}\} = [\tilde{B}] \{\tilde{u}\}. \tag{31}$$

The objective is to estimate the contribution of the *convective stiffness matrix* arising out of the additional quantities in system stochasticity in a problem independent uniform manner. In a single element domain Ω, the conventional deterministic static stiffness matrix, using the deterministic constitutive matrix $[D]$ is given by:

$$[K] = \int_{\Omega} [B]^{\mathscr{T}} [D] [B] \, d\Omega. \tag{32}$$

Likewise, the stochastic stiffness matrix should become:

$$[\tilde{K}] = \int_{\Omega} [\tilde{B}]^{\mathscr{T}} [\tilde{D}] [\tilde{B}] \, d\Omega \tag{33}$$

in terms of the sample constitutive field $[\tilde{D}]$.

It is necessary to point out that in the *ad hoc* approximate procedure of using the conventional element Subroutine without considering stochastic shape functions, the approximate stochastic stiffness matrix $[\mathscr{K}^{\text{app.}}]$, using the deterministic shape functions with material stochasticity, will be none other than:

$$[\mathscr{K}^{\text{app.}}] = \int_{\Omega} [B]^{\mathscr{T}} [\tilde{D}] [B] \, d\Omega. \tag{34}$$

The (hybrid) stress equilibrium statement is given according to Ref. [4] by:

$$[\tilde{D}] [\tilde{B}] = [D] [B]. \tag{35}$$

This ensures the same order of error for all randomly simulated samples as far as the stress distributions are concerned. Essentially, the stress profile relationship provides a direct evaluation of the stochastic strain-displacement matrix $[\tilde{B}]$ in the form:

$$[\tilde{B}] = [\tilde{D}]^{-1} [D] [B]. \tag{36}$$

This stochastic strain-displacement transformation, Eq. (31), is recommended to be used in stochastic finite element computer programs. In order to keep as much as (say Fortran) computer statements from the deterministic code as possible an additional convective matrix $[K^{\mathscr{C}}]$ is to be defined by:

$$[K^{\mathscr{C}}] = \int_{\Omega} [B]^{\mathscr{T}} [D^{\mathscr{C}}] [B] \, d\Omega \quad \text{such that} \quad [\tilde{K}] = [K^{\text{app.}}] + [K^{\mathscr{C}}]. \tag{37}$$

The "equivalent convective" constitutive matrix to be employed is then given by:

$$[D^{\mathscr{C}}] = [D][\tilde{D}]^{-1}[D] - [\tilde{D}]. \tag{38}$$

The aforementioned equation is easy to code. It has been incorporated in an existing deterministic finite element computer program to generate the dynamic responses presented here. No additional complexity is encountered either in the time or in the frequency domain calculations.

5 Numerical examples

In the case of mild stochasticity, the perturbation method, [5] and [10], is adequate to obtain the response statistic for dynamic problems. Hence, in this paper the critical problem of large random variability is focused. Since the perturbation diverges for large stochastic dispersions the Monte Carlo method, extensively employed by Shinozuka [23], seems to be the only available numerical tool which could be deployed in the existing computing environment [24]. For each sample the solution is obtained according to the stationary iteration scheme initiated by Boley in [1] and [2] and accelerated according to [13].

During the numerical computations, representative sample populations, ranging from one hundred to one thousand, were generated with exponentially decaying "homogeneous" covariance functions of the form:

$$\text{Cov}\,[E(r),\,E(r+r')] = \exp\left(-c_o\frac{r'}{l}\right) \tag{39}$$

where r' is the distance between two points r and $r+r'$, l is a characteristic linear dimension, and c_o is a nondimensional factor. The aspect of generation of bounded random fields [22] for constitutive properties during those Monte Carlo simulations is somewhat extraneous to the principal theme of dynamic calculations involving stochastic fields, hence, left out. Two representative spatial distributions of modulus of elasticity are depicted in Figs. 1 and 2, respectively, for rapidly and slowly dying correlation effects. In the latter, twenty five samples, generated according to the principal component analysis, are shown.

The dynamic problems solved for stochastic material properties are carried out both in the frequency and time domains. It is to be noticed that the stochastic aspects basically capture the secondary random effects. In dynamic discrete calculations there are many conjectures regarding how to combine different proportions of consistent and lumped masses. Hence, only the two extremes, i.e. fully consistent and purely lumped masses were entertained during dynamic simulations presented here. For these two classes of mass matrices, negligible amount of stochastic deviations were found both in the time and frequency domain schemes.

The time harmonic formulations, Eqs. (12) through (19), were employed in boundary element calculations. Keller's formula for the mean field, Eq. (19), as suggested by its rigorous deviation, was verified to be extemely accurate. The numerical treatment was carried out for beam and membrane problems. The Green's fundamental solutions were generated in algebraic form using Mathematica [25]. These functions were then linearly combined to generate displacement shape functions in the usual fashion. These interpolants were input in dynamic stiffness generation routines. The mode shapes for the beam problems, for the first and third harmonics are presented in Figs. 3 and 4. The statistics for the frequencies are computed in a straightforward fashion from

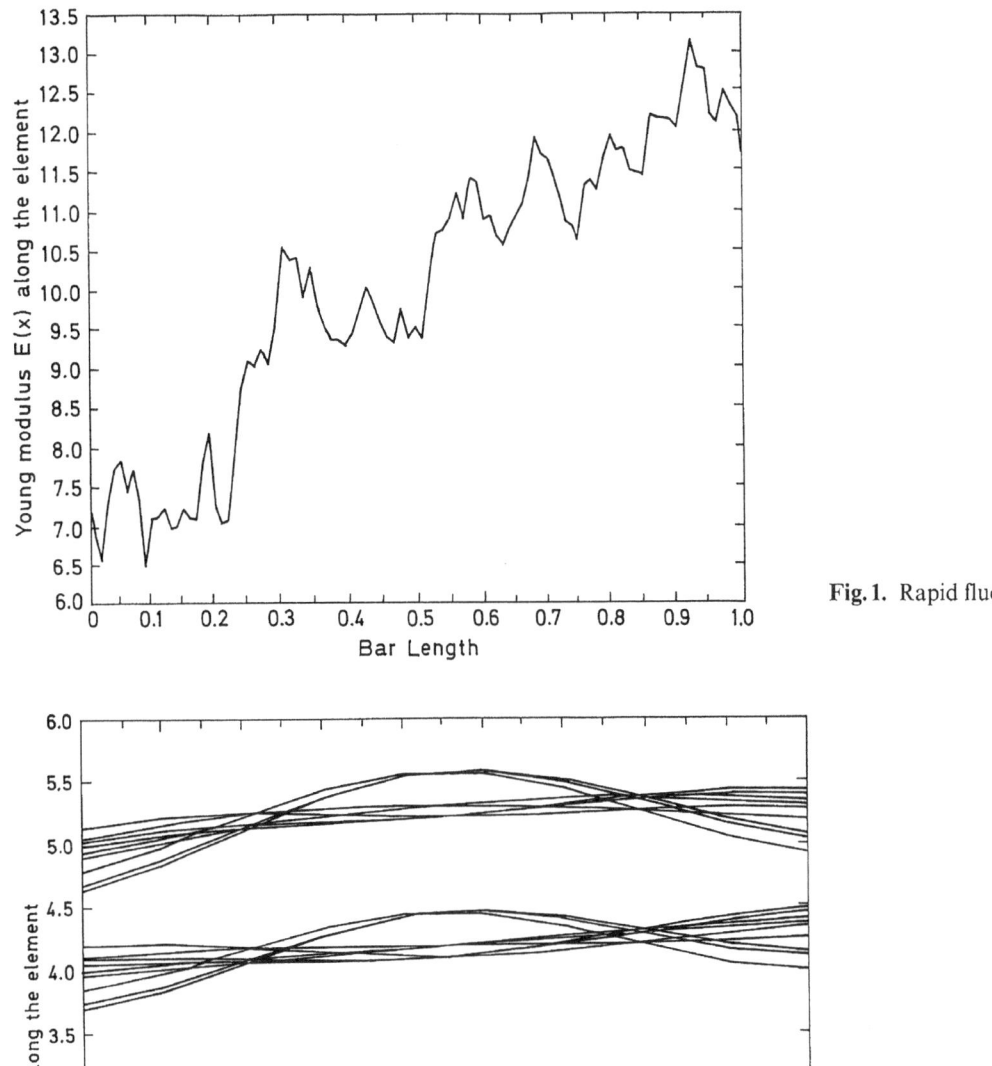

Fig. 1. Rapid fluctuation

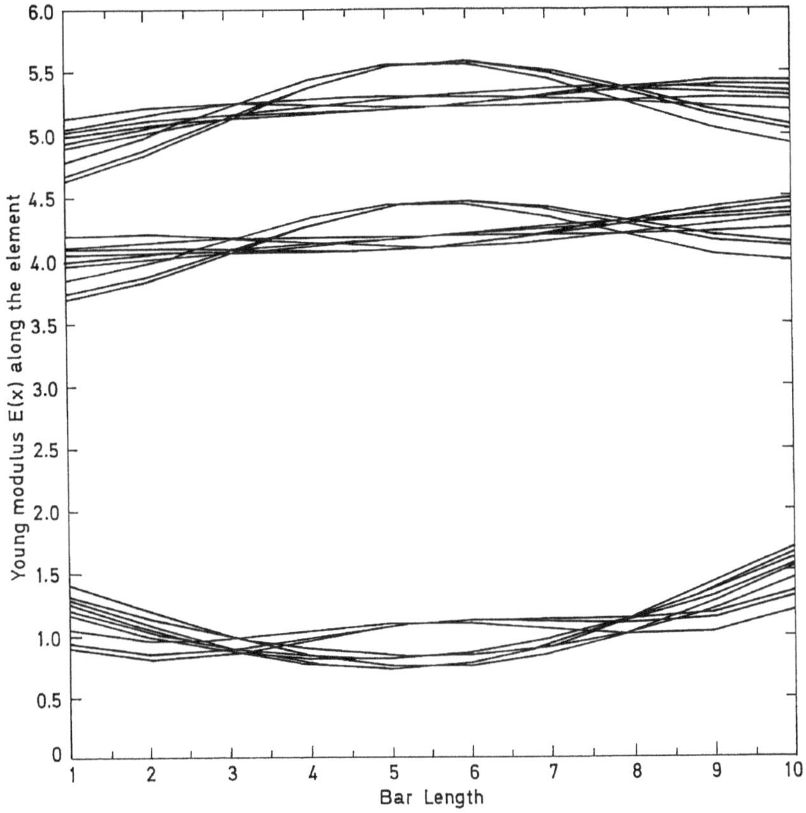

Fig. 2. Mild variability

individual samples. For the membrane problem similar responses were obtained. The computed mean field for the first four harmonics were compared with Eq. (19). The agreement was remarkable, as anticipated.

For an axially loaded bar, the analytical expression for exponentially decaying spatial correlation of the modulus of elasticity was assumed in the form of Eq. (39). A critical simulated

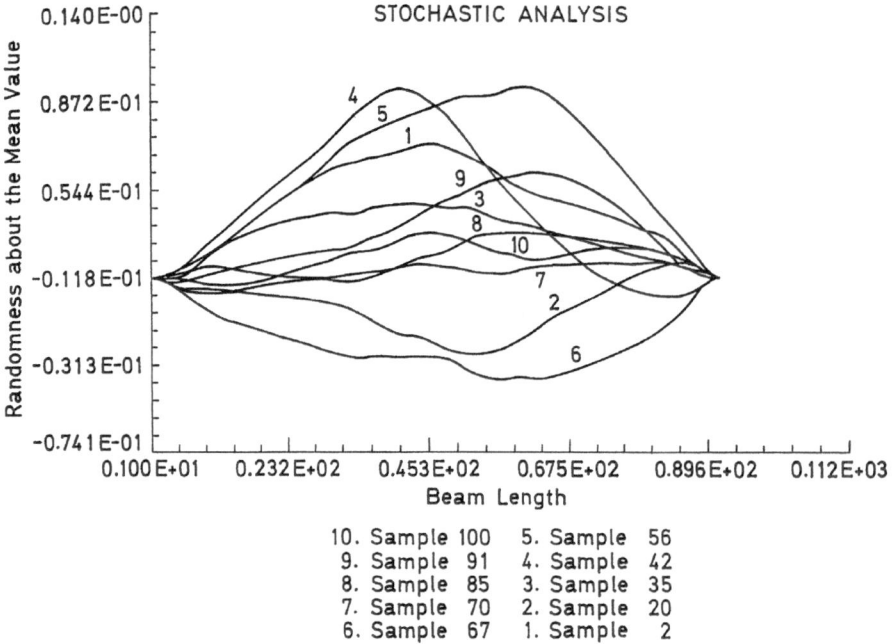

Fig. 3. Stochastic deviation of the first mode

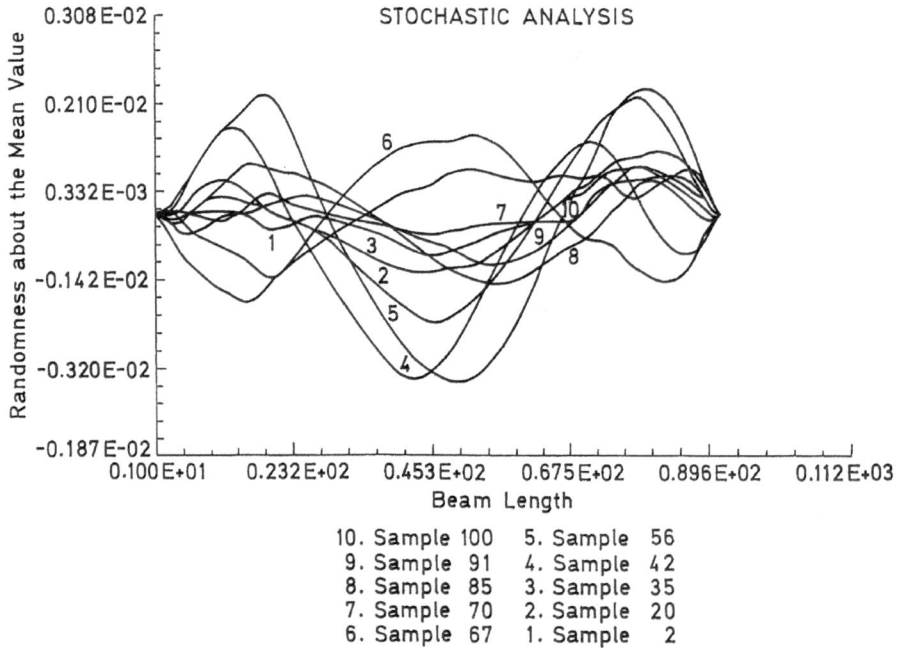

Fig. 4. Stochastic deviation of the third mode

sample can be found in Fig. 1. The effect of randomness in the *stochasic shape function* is presented in Fig. 5. The analytical results for the end displacements, computed in closed algebraic form, with Mathematica, [25], and was verified according to the Monte Carlo simulation technique.

With the aforementioned correlated modulus of elasticity stochastic finite elements were generated for beams. For the purpose of dynamic response calculations with impact loading, the lumped mass matrix was employed. Hence, the mass matrix remained deterministic, whereas, the

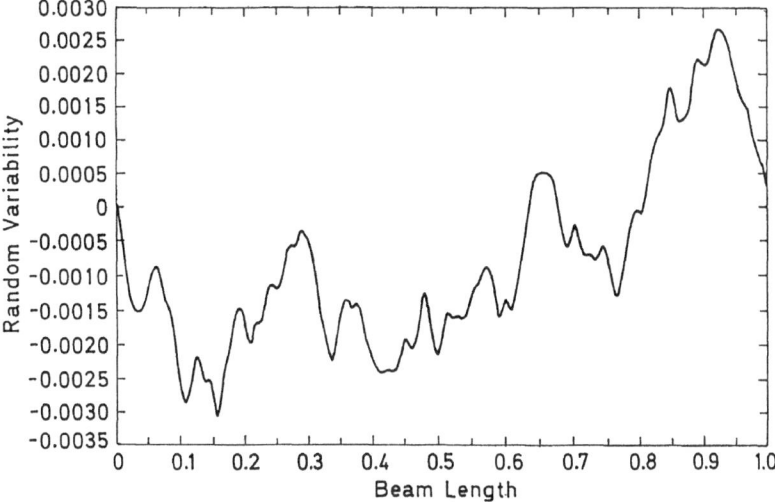

Fig. 5. Fluctuation of a shape function

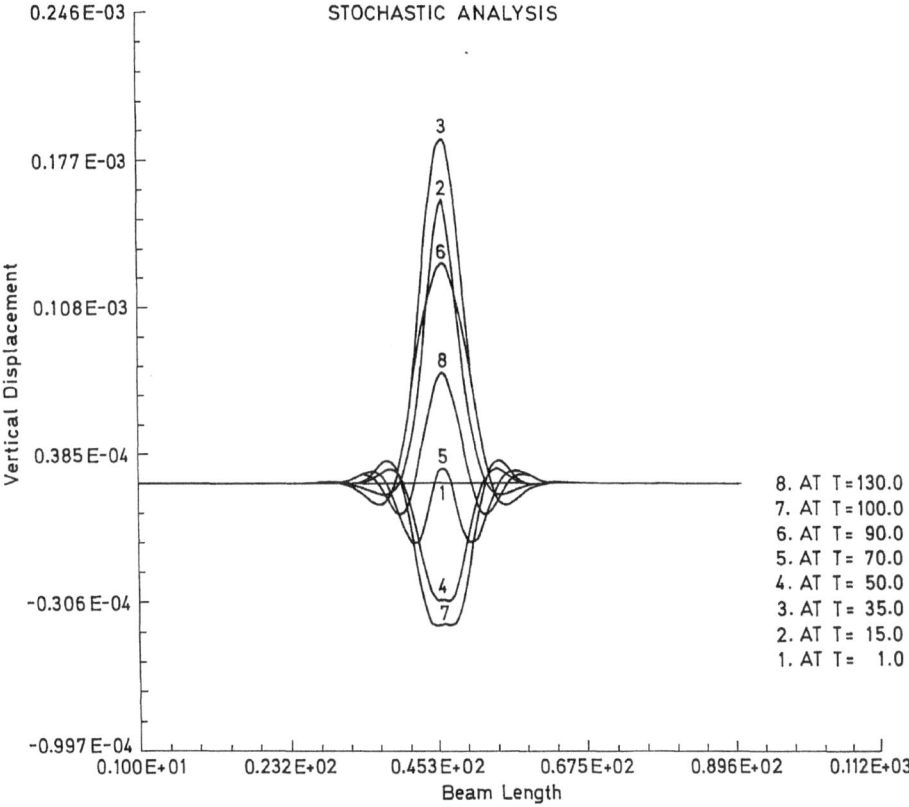

Fig. 6. A typical displacement history

stiffness matrix was modeled as a random quantity. In order to investigate the extent of randomness due to material stochasticity, the convective component [7] of the "constitutive matrix" stated in Eq. (38), was calculated. The consequences of ignoring the convective terms can be found in [26]. The same deterministic Subroutine was employed to generate the approximate stiffness matrix, Eq. (37), in the first pass. Next, the additional convective part was calculated calling the same Subroutine. Different simulated samples are shown with numerical tags in Fig. 4. The mean dynamic response to an impact loading is presented in Figs. 6 and 7. Eight

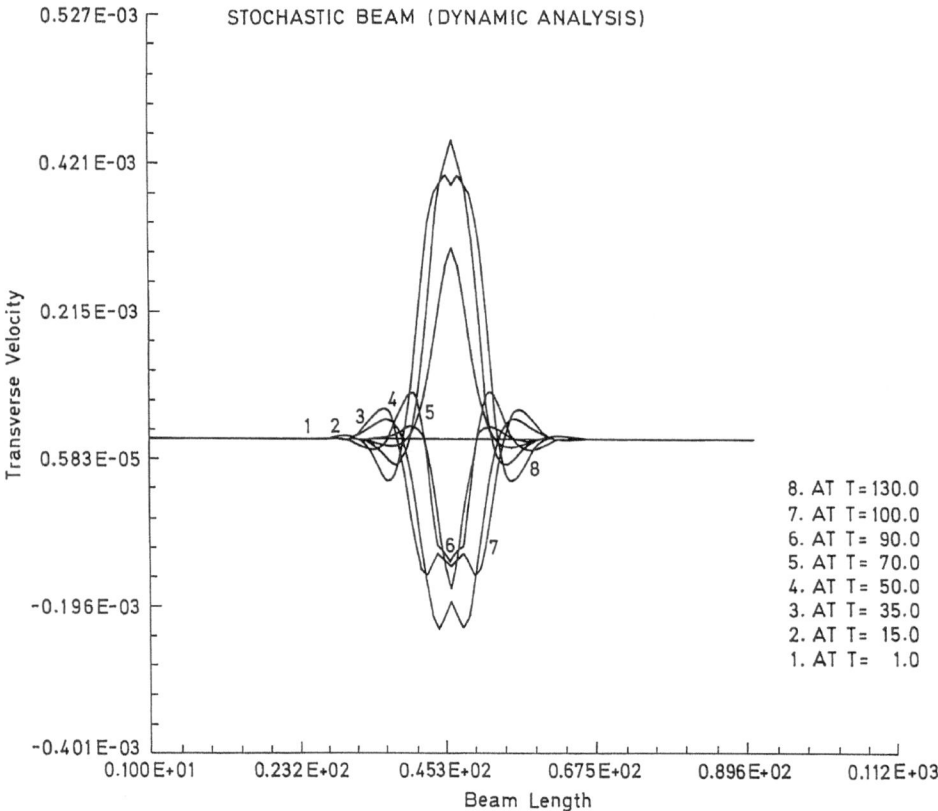

Fig. 7. A typical velocity history

responses at different time T are identified. The convective terms for the static stiffness were examined to contribute almost 30% in the velocity correlation during the Monte Carlo simulation. In general, statistical stability was attained with 100 samples for the dynamic beam problem.

6 Conclusions

Stochastic formulations to compute dynamic responses furnish the response statistics suitable for engineering design-analysis. The probabilistic feature of the dynamic operator adds an interesting dimension to computational issues. The complexity of the stochastic algorithms are drastically reduced by the proposed stationary iterative strategy, which is styled after Boley [1]. Therein, the primary solutions of the thermo-elastodynamic sytem are computed using the uniform constitutive parameters considering uncoupled problems. Large fluctuations in the material properties are incorporated in the subsequent samples.

For wave propagation problems, and for the steady harmonic responses as well, the lumped mass matrix representation was found adequate to detect the random fluctuations about the mean values. This simplification was helpful to keep the focus on *stochastic shape functions*. The possibility of transforming a system stochasticity problem into a random loading one has been examined here in depth. It was found adequate to study the one-dimensional finite element analysis for a shear beam and subsequently the flexural one. First, the concept of equivalent random loading formulation, for simple non self-adjoint *system stochasticity* problems, was demonstrated. The need to re-derive the stochastic differential equation for self-adjoint cases was then addressed to conclude the consequences from a representative statement, e.g. Eq. (29). Shortcomings of extending a deterministic computer program in stochastic finite element formulation for dynamic problems, by simply interchanging the deterministic constitutive matrix by its stochastic counterpart, were detected and were emphasized in the numerical coding. It is established that the output correlation of a self-adjoint field equation would be deficient if the contribution of convective type terms are ignored. Stochasticity for finite element interpolants, similar to those which are routinely accounted for in modeling fluid turbulence [7] and in gas dynamics problems [5] and [8], is recognized to be important. In reliability analysis of dynamic structures these additional convective terms are of primary interest.

Keller's idea, [15] and [16], to construct the operator equation for the mean field is incorporated in the proposed stochastic boundary element and subsequently into the stochastic finite element formulation. The smoothing assumption permits the flux field to be deterministic for a set of deterministic boundary nodal displacements. Thus an exorbitant computational cost of re-meshing for each simlated fluctuation is avoided. Now, the Monte Carlo simulation procedure becomes very effective.

The proposed methodology exhibits a synthesis of stochastic dynamics and Boley's papers, [1] and [2] in the boundary element applications. For the finite element method the technique that combines [1] and [7] of Tatarski is presented here. The present formulation should be applicable to unbounded media as well. For example, the inhomogeneous random soil media can be incorporated in foundation-structure interactions, [3]. Currently, for such time dependent problems, computer algebra softwares are being implemented. This part of the computation is under development using *MathLink* in Mathematica [25]. Attempts are underway also to model the layered stochastic media.

Acknowledgment

The author expresses his gratitude for the help rendered by Mr. J. Ueno, Konoike Construction, Osaka, Japan and Prof. M. E. McAlarney, School of Dentistry and Oral Surgery, Columbia University, New York, NY, while preparing this paper. Research support from the National Science Foundation, USA and Humboldt Foundation, Bonn, Germany, is gratefully cited.

References

[1] Boley, B. A.: A method for the construction of Green's Functions. Quart. Appl. Math. **14**, 249−257 (1956).

[2] Boley, B. A.: A method for the construction of fundamental solutions in elasticity theory. J. Math. Phys. **36**, 261−268 (1957).

[3] Dasgupta, G.: Stochastic finite element analysis of soil-structure systems, Vol. II, pp. 528−532, Fourth International Conference on Structural Safety and Reliability, Kobe, Japan, 1985.

[4] Dasgupta, G.: Green's functions for inhomogeneous media for boundary elements. In: Advances in boundary elements (Brebbia, C. A., Connor, J. J., eds.), pp. 37−46, Ashurst, Southampton, UK: Computational Mechanics Publications 1989.

[5] Sobczyk, K.: Stochastic Wave Propagation. New York: Elsevier 1985.

[6] Lee, X., Dasgupta, G.: Analysis of structural variability with computer algebra. J. Eng. Mech. **144**/1, 161−171 (1988).

[7] Tatarski, V. I.: Wave Propagation in a Turbulent Medium. New York: McGraw-Hill 1961.

[8] The Proceedings of Converence on Sonic Boom Research, NASA Sp. Rpt. SP-180, May 9−10, 1968.

[9] Astill, C. J., Nosseir, B., Shinozuka, M.: Impact loading on structures with random properties. J. Struct. Mech. **1**, 63−77 (1972).

[10] Nakagiri, S., Hisada, T.: Notes on stochastic finite element method. Tokyo, Japan: Seisan-Kenkyu, research journal of the institute of industrial sciences, Vols. 32−35 (in eight parts), 1979.

[11] Nakagiri, S., Hisada, T.: An introduction to the stochastic finite element method. Tokyo, Japan: Baifu-Kan 1985 (in Japanese).

[12] Dasgupta, G., Yip, S-C.: Nondeterministic shape functions for finite elements with stochastic moduli. Vol. II, pp. 1065−1072, 5th International Conference on Structural Safety and Reliability, San Francisco, CA, 1989.

[13] Dasgupta, G.: Simulation with ordered discrete-stochastic samples. In: Probabilistic Methods in Civil Engineering (Spanos, P. D., eds.), pp. 148−151, New York, NY: ASCE publication 1988.

[14] Stochastic differential equations. American Mathematical Society, Providence, RI, Vol. VI, SIAM-AMS Proceedings (Keller, J. B., McKean, H. P., eds.) 1973.

[15] Keller, J. B.: Wave propagation in random media. In: Symposium on applied mathematics, No. 13, pp. 227−246, Providence, RI: American Mathematical Society 1962.

[16] Keller, J. B.: Stochastic equations and wave propagation in random media. In: Symposium on Applied Mathematics, No. 16, pp. 145−170, Providence, RI, American Mathematical Society 1964.

[17] Brebbia, C. A., Dominguez, J.: The Boundary Element Method. Ashurst, Southampton, UK: Computational Mechanics Publications 1989.

[18] BEASY: User's manuals, Ashurst, Southampton, UK: Computational Mechanics Publications 1991.

[19] Beskos, D. E.: Dynamic analysis of beams, plates and shells, ls. In: BEM in structural analysis, pp. 139−161 (Beskos, D. E., ed.), New York, NY: ASCE Publication 1989.

[20] Camp, C. V., Gipson, G. S.: Boundary element analysis of nonhomogeneous biharmonic phenomena. Ashurst, Southampton, UK: Computational Mechanics Publications 1990.

[21] Vanmarcke, E., Grigoriu, M.: Stochastic finite element analysis of simple beams. J. Eng. Mech., ASCE. **109**, 1203−1214 (1983).

[22] Yamazaki, F., Shinozuka, M.: Digital generation of non-gaussian stochastic fields, Stochastic Mechanics, Vol. I, (Shinozuka, M., ed.), Department of Civil Engineering and Engineering Mechanics, Columbia University 1986.

[23] Shinozuka, M.: Monte Carlo solution of structural dynamics. International J. Comput. Struct. **2**, 855−874 (1972).

[24] Shinozuka, M., Dasgupta, G.: Stochastic finite element methods in dynamics, Keynote Paper, Proceedings, ASCE Specialty Conference on the Dynamic Responses of Structures, pp. 44−54, 1986.

[25] Wolfram, S.: Mathematica − A system for doing mathematics by computer, 2nd ed., Redwood City, CA, Addison-Wesley 1991.

[26] Molyneux, J. E.: Analysis of "Dishonest" methods in the theory of wave propagation in a random medium. J. of the Optical Society of America. **58**, 951−957 (1968).

Author's address: Prof. G. Dasgupta, Department of Civil Engineering and Engineering Mechanics, Columbia University, New York, NY 10027-6699, U.S.A.

Acta Mechanica (1992) [Suppl] 3: 115–127

Self-excited vibrations of a flexible disk rotating on an air film above a flat surface

H. Hosaka, Tokyo, Japan, and S. H. Crandall, Cambridge, Massachusetts, U.S.A.

Summary. Self-excited vibration of a spinning flexible disk is studied to provide design guidance for high speed magnetic disk storage devices. The dynamic model includes coupling between vibrations of a rotating plate and waves in an incompressible viscous fluid film. A single-mode approximation is used to estimate the critical speed and the influence of the various disk and air-film parameters is noted. A simple physical explanation of the instability mechanism is given and the critical speeds for several disk configurations are estimated. More accurate numerical procedures are described and the results of calculations for a particular disk system are compared with the single-mode approximation.

1 Introduction

Flexible disks are widely used as a low-cost recording medium for computer memories. The recent drastic increase in the quantity of computer data has led to a strong demand for high-speed, high-reliability devices based on flexible disks. The gas lubricated flexible disk drive was invented to meet this requirement (Pearson, 1961). In this drive, a disk rotates close to a rigid plate, and a self-acting air bearing is created between the disk and the plate. To design this system, therefore, it is necessary to analyze the interaction between the disk deflection and the air flow.

Pelech and Shapiro (1964), Bogy and Talke (1978), Adams (1980), Licari and King (1981), and Hosaka and Nishida (1987) studied steady-state characteristics of the disk-air system. For unsteady conditions, Sato (1985) studied damped vibration, Ono and Maeno (1986) studied forced vibration, and Adams (1987) studied critical speed of resonance. However unstable vibration has not yet been studied even though its prediction is the most important task in the design of any rotating mechanism.

In this paper self-excited vibration of the system is studied. A single-mode analysis is performed to demonstrate the existence of instability and to obtain a rough estimate of the critical speed. The nature of the instability mechanism is examined and a simple physical explanation is given. Finally more accurate numerical analyses are introduced and results obtained which provide a basis for judging the accuracy of the single-mode analysis.

2 Mathematical model

A simplified model of a flexible disk recording device is shown in Fig. 1. A flexible disk spins over a rigid base plate at an angular velocity Ω. A thin air film generated between the disk and the plate acts as a fluid-film bearing. To analyze the dynamics of this system, the following assumptions are introduced:

— Air film thickness is small and air flow is laminar.
— Air flow on top of the disk does not affect disk deflection.

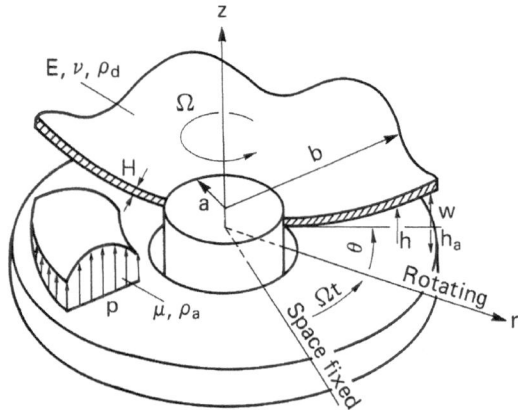

Fig. 1. Analytical model of flexible disk drive

- Disk vibration amplitude is much smaller than the air film thickness.
- Air inertia contributes only as centrifugal force.
- Disk material is viscoelastic with internal damping force proportional to the strain velocity.

The governing equations are a thin-plate bending equation for the transverse deflection w driven by the air-film pressure p and a Reynold's equation for p driven by the speed Ω and by the varying film thickness $h = h_a + w$ (h_a is the hub height). Polar coordinates r, θ with respect to axes fixed in the rotating disk are employed and centrifugal loading is accounted for by including the centrifugal stress components σ_r, σ_θ in the plate bending equation and by including a term proportional to Ω^2 in Reynold's equation (Pinkus and Lund, 1981). The (nonlinear) governing equations are

$$\varrho_d H \frac{\partial^2 w}{\partial t^2} + \eta D \frac{\partial}{\partial t} (\nabla^4 w) + D\nabla^4 w - \frac{H\partial}{r \, \partial r} \left(\sigma_r r \frac{\partial w}{\partial r} \right) - \frac{H}{r^2} \sigma_\theta \frac{\partial^2 w}{\partial \theta^2} = p \tag{1}$$

$$\frac{\partial}{\partial r} \left(rh^3 \frac{\partial p}{\partial r} \right) + \frac{\partial}{r \, \partial \theta} \left(h^3 \frac{\partial p}{\partial \theta} \right) = 12\mu r \frac{\partial h}{\partial t} - 6\mu\Omega r \frac{\partial h}{\partial \theta} + \frac{3}{10} \varrho_a \Omega^2 \frac{\partial}{\partial r} (r^2 h^3) \tag{2}$$

where H is the disk thickness, $D = EH^3/12(1 - v^2)$ is its flexible rigidity, ϱ_d and ϱ_a are disk and air densities, η is the viscoelastic damping coefficient for the disk, and μ is the viscosity of the air. The boundary conditions for (1) are those of a clamped edge at $r = a$ and a free edge at $r = b$. For (2) the film pressure equals the ambient pressure at both the inner and outer edges.

The steady-state deflection w_0 and pressure p_0 are calculated first. By setting all t-derivatives and all θ-derivatives in (1) and (2) to zero, ordinary differential equations for $w_0(r)$ and $p_0(r)$ are obtained (Adams, 1980). Here, where only high-speed spin is considered, the bending stiffness can be neglected with respect to the dominant membrane stiffness in determining the steady-state configuration. In this approximation (1) reduces to a second-order differential equation and w_0 and p_0/Ω^2 become independent of Ω since both the membrane stiffness and the air-film centrifugal force are proportional to Ω^2. A numerical procedure for solving this problem is given by Pelech and Shapiro (1964).

Next, linearized dynamic equations in the neighborhood of the steady state are obtained by setting $w = w_0 + w'$ and $p = p_0 + p'$ and neglecting higher order terms in the small perturbations w' and p'. With the assumption of an n nodal diameter mode shape

$$w' = w_r(r) \, e^{in\theta} e^{\lambda t} \tag{3}$$

$$p' = p_r(r) \, e^{in\theta} e^{\lambda t} \tag{4}$$

the governing equations reduce to the pair of ordinary differential equations

$$\lambda^2 \varrho_d H w_r + \lambda \eta K_b[w_r] + K_b[w_r] + K_m[w_r] = p_r \tag{5}$$

$$R[p_r] + N[w_r] + (\lambda - in\Omega/2)\, w_r = 0 \tag{6}$$

where

$$K_b[w_r] = D\left(\frac{d^4}{dr^4} + \frac{2}{r}\frac{d^3}{dr^3} - \frac{2n^2+1}{r^2}\frac{d^2}{dr^2} + \frac{2n^2+1}{r^3}\frac{d}{dr} + \frac{n^4-4n^2}{r^4}\right) w_r$$

$$K_m[w_r] = -\frac{H}{r}\frac{d}{dr}\left(r\sigma_r \frac{dw_r}{dr}\right) + \frac{Hn^2}{r^2}\sigma_0 w_r$$

$$\tag{7}$$

$$R[p_r] = \frac{1}{12\mu r}\left[\frac{n^2 h_0{}^3}{r}\, p_r - \frac{d}{dr}\left(r h_0{}^3 \frac{dp_r}{dr}\right)\right]$$

$$N[w_r] = \frac{1}{4\mu r}\left[\frac{3}{10}\varrho_a\Omega^2 \frac{d}{dr}(r^2 h_0{}^2 w_r) - \frac{d}{dr}\left(r h_0{}^2 \frac{dp_0}{dr}\, w_r\right)\right]$$

and $h_0 = h_a + w_0$ is the steady state air-film thickness.

3 Single-mode analysis

3.1 Eigenvalue problem; onset of instability

Equations (5) and (6) can be solved numerically as shown in § 4. Here, however, a simplified analytical method is used to obtain a rough estimate of the critical speed. First, we put

$$w_r(r) = W\phi(r)$$

$$\tag{8}$$

$$p_r(r) = P\psi(r)$$

in (5) and (6), where ϕ and ψ are trial functions, and W and P are amplitude parameters. For ϕ, we use the membrane mode function (Lamb and Southwell, 1921),

$$\phi = (r/b)^n \tag{9}$$

which is expected to be the simplest and the most reasonable approximation when a is small. For ψ, we use the following function

$$\psi = \phi(b-r)/b \tag{10}$$

which is also simple and satisfies the boundary condition for p_r at $r = b$. Then (5) is multiplied by $r\phi$ and (6) is multiplied by $r\psi$ and both equations are integrated from $r = a$ to $r = b$ to obtain a pair of homogeneous linear algebraic equations for W and P with eigenvalue λ. The characteristic equation for nontrivial solutions is

$$\varrho_d H\lambda^2 + \eta k_b \lambda + c(\lambda - in\Omega/2) + k_b + k_m + k_a = 0 \tag{11}$$

where in terms of the inner product notation $\langle f, g\rangle = \int_a^b fgr\, dr$ the coefficients are

$$k_b = \langle K_b[\phi],\, \phi\rangle / \langle\phi,\, \phi\rangle$$

$$k_m = \langle K_m[\phi],\, \phi\rangle / \langle\phi,\, \phi\rangle$$

$$\tag{12}$$

$$k_a = \langle\phi,\, \psi\rangle\, \langle\psi,\, N[\phi]\rangle / \langle\phi,\, \phi\rangle\, \langle\psi,\, R[\psi]\rangle$$

$$c = -2\langle\phi,\, \psi\rangle^2 / \langle\phi,\, \phi\rangle\, \langle\psi,\, R[\psi]\rangle.$$

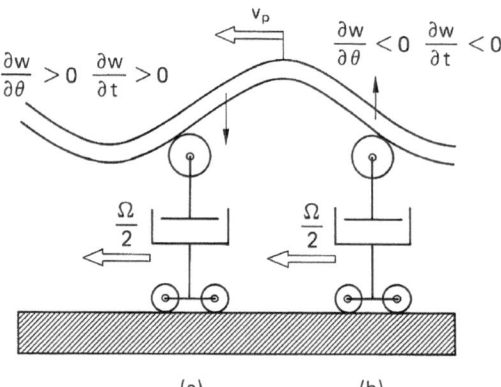

$$\frac{\partial w}{\partial \theta} > 0 \quad \frac{\partial w}{\partial t} > 0 \qquad \qquad \frac{\partial w}{\partial \theta} < 0 \quad \frac{\partial w}{\partial t} < 0$$

(a)   (b)

Fig. 2. Beam-dashpot model of flexible disk system

Equation (11) has the same form as the characteristic equation for wave propagation with wave number n on a one dimensional continuous beam of mass per unit length $\varrho_d H$ with bending stiffness k_b, membrane stiffness k_m, and foundation stiffness k_a when the beam is propelled longitudinally with velocity Ω and is subjected to two kinds of damping. One is internal damping proportional to bending strain (with coefficient η) and the other is external damping proportional to transverse velocity (with coefficient c) applied by a system of dashpots which are themselves propelled longitudinally with velocity $\Omega/2$. In such a system the external damping is destabilizing for backward waves whose phase velocity v_P is smaller than $\Omega/2$. This can be seen in Fig. 2 where the continuum is viewed from a coordinate system moving at velocity Ω. The external dashpots appear to be moving backwards with velocity $\Omega/2$ and the bending wave has a phase velocity v_P. When $v_P > \Omega/2$ we have the case shown in Fig. 2. At (a) where $\partial w/\partial \theta > 0$ the dashpot pulls down on a point of the beam that is moving up ($\partial w/\partial t > 0$). At (b) where $\partial w/\partial \theta$ the dashput pushes up on a point of the beam where the beam is moving down ($\partial w/\partial t < 0$). The dashpot forces oppose the local transverse velocity and remove energy from the wave. The steady motion is stabilized. When $\Omega/2 > v_P$, however, the situation is reversed. The dashpot forces add energy to the wave which tends to destabilize the steady motion. In a system with internal damping the onset of instability occurs when the destabilizing power input of the external damper first neutralizes the power loss due to internal damping. To obtain the critical speed for onset of instability in (11) we use the fact that at the stability borderline the eigenvalue will be purely imaginary. When $\lambda = i\omega$ is inserted in (3), (4) we find $v_P = w/n$ and when inserted in (11) the separate vanishing of the real and imaginary parts yields

$$\omega^2 = \frac{k_b + k_m + k_a}{\varrho_d H} \quad \text{and} \quad \frac{\Omega cr}{2} = v_P \left(1 + \frac{\eta k_b}{c} \right). \tag{13}$$

Alternatively, the Bilhartz criteria (Bilhartz, 1944) can be used to show that the λ-roots of (11) lie in the left half-plane when $\Omega < \Omega_{cr}$. The instability mechanism here has much in common with that for a centered journal in a continuous fluid film bearing. There, a rotor whirl at frequency ω is destabilized when the rotation rate Ω satisfies the relation $\Omega/2 > \omega$ (Robertson, 1935). In both cases there is Couette flow in the circumferential direction and the average circumferential velocity in the fluid film is $\Omega/2$ times the appropriate radius. The manner in which the stability onset speed is determined here by the interaction of a negative damping element with a positive damping element is similar to that for unstable shaft whirl due to destabilizing damping in the rotor and stabilizing external drag on the rotor (Smith, 1936) with the exception that here the roles of internal and external damping are interchanged.

3.2 Estimation of parameter values

To estimate the air-film stiffness k_a and damping coefficient c it is useful to rewrite the terms $\langle \psi, N[\phi] \rangle$ and $\langle \psi, R[\phi] \rangle$ in (12). When the relationship between p_0 and h_0 obtained from (2) is inserted in $N[\phi]$ we obtain the alternative representation

$$\langle \psi, N[\phi] \rangle = -\frac{3}{80} \frac{\varrho_a \Omega^2 \bar{h}_0^3}{\mu} \frac{b^2 - a^2}{\ln (b/a)} \left\langle \frac{\psi}{r}, \frac{d}{dr} \left(\frac{\phi}{h_0} \right) \right\rangle \tag{14}$$

where $\bar{h}_0$ is a weighted average air-film thickness given by

$$\int_a^b \frac{dr}{r h_0^3} = \frac{1}{\bar{h}_0^3} \int_a^b \frac{dr}{r} = \frac{1}{\bar{h}_0^3} \ln (b/a). \tag{15}$$

After integration by parts the term $\langle \psi, R[\psi] \rangle$ reduces to

$$\langle \psi, R[\psi] \rangle = \frac{1}{12\mu} \int_a^b h_0^3 \left[r \left(\frac{d\psi}{dr} \right)^2 + \frac{n^2}{r} \psi^2 \right] dr. \tag{16}$$

To obtain rough approximations to $\langle \psi, N[\phi] \rangle$ and $\langle \psi, R[\psi] \rangle$ which have $h_0(r)$ in the integrand we use the fact that ϕ, ψ, and their derivatives are only large in the neighborhood of $r = b$, to replace the variable $h_0(r)$ by the constant $h_0(b)$. In this way we obtain the approximations

$$k_a \approx \varrho_a \Omega^2 H_a \quad \text{with} \quad H_a = \frac{9}{20} \frac{b^2 - a^2}{h_a \ln b/a} \frac{\beta^3}{\alpha^4} \frac{n + 1}{(2n + 1)(2n + 3)}$$

$$c \approx \frac{12\mu b^2}{h_a^3} \frac{1}{\alpha^3 (2n + 3)^2} \tag{17}$$

where

$$\alpha = \frac{h_0(b)}{h_a}, \quad \beta = \frac{\bar{h}}{h_a}. \tag{18}$$

Since h_0 is substantially independent of Ω in the high-speed region, we can regard α and β as constants.

Next we study the bending stiffness k_b. It is found that $k_b \equiv 0$ if we approximate ϕ by the membrane mode function given by (9). The estimation of k_b is crucial for the study of system stability as will be shown later. It is therefore necessary to estimate it more precisely. Since k_b has the form of a Rayleigh quotient we can estimate it from the known natural frequency of the stationary plate vibration. The parameter k_b thus has the value

$$k_b = \chi_n^4 D/b^4 \tag{19}$$

where the dimensionless parameter χ_n has been tabulated (Itao and Crandall, 1979). For small values of a/b and moderately large values of n, χ_n is roughly approximated by n.

The membrane stiffness k_m is directly obtained by inserting (9) in (12), and has the value

$$k_m = \varrho_d H \Omega^2 [(1 - \nu) n^2 + (3 + \nu) n]/4 \tag{20}$$

when a/b is assumed to be small.

By inserting (17) through (20) into (13) and solving for Ω^2 at the onset of instability we obtain

$$\Omega_{cr}^2 = \frac{EH^2\chi_n{}^4}{3\varrho_d b^4(1 - v^2)\,[(v - \zeta)\,n^2 - (3 + v)\,n - 4\varrho_a H_a/\varrho_d H]} \tag{21}$$

for the n-nodal diameter mode, where

$$\zeta = 1 - \frac{1}{(1 + \eta k_b/c)^2} \tag{22}$$

is a measure of the importance of the internal damping in comparison with the fluid-film damping. Since ζ and H_a are non-negative, the modes for low values of n are stable. The critical speed is not real until the right hand side of (21) is positive which requires light damping ($\zeta < v$) and high values of n. A lower bound for the n-value of the first unstable mode is

$$n > 1 + \frac{3}{v}. \tag{23}$$

3.3 Effect of system parameters on critical speed

The effect of each system parameter on the critical speed at the onset of instability for a given mode can be predicted from a careful study of the phase velocity v_P of the circumferential wave in the disk and the effective relative transport velocity of the air-film dashpot-system. It is helpful to consider first the special case where internal damping is negligible ($\zeta \to 0$) and the air-film stiffness is negligible ($k_a \to 0$). In this case the phase velocity v_P is given by

$$v_P{}^2 = \frac{k_b + k_m}{n^2\varrho_d H} \tag{24}$$

while the effective relative transport velocity v_d of the air-film damper is $\Omega/2$, or

$$v_d{}^2 = \Omega^2/4. \tag{25}$$

Equations (24) and (25) are displayed in Fig. 3. The ordinate of $v_P{}^2$ is decomposed into a bending contribution, independent of spin speed, and a centrifugally loaded membrane stiffness contribution, proportional to Ω^2. The critical speed is determined by the intersection of (24) and (25). The stable operating range is extended (Ω_{cr} is increased) if the bending stiffness contribution

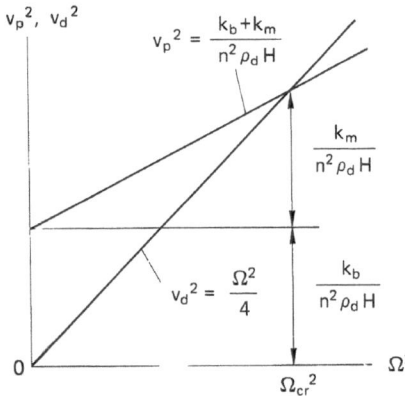

Fig. 3. Phase velocity v_p of disk bending wave, and dashpot transport velocity v_d of average air flow, as functions of rotational speed Ω

is increased; i.e., if E, H, or n is increased or if ϱ_d or b is decreased. The membrane stiffness contribution, on the other hand, is increased by a decrease in n. The critical speed has its minimum value for a certain n which can be obtain by minimizing (21) with respect to n. With $\chi_n = n$ we find

$$\Omega_{\min} = \frac{3H(3+v)}{2b^2}\left(\frac{E}{v^3(1-v^2)\,\varrho_d}\right)^{1/2} \tag{26}$$

for

$$n = \frac{3(3+v)}{2v}. \tag{27}$$

The effect of including internal damping ($\eta > 0$) can be deduced by rewriting the second and third terms of (11) as

$$(c + \eta k_b)\{\lambda - in[\Omega/2(1 + \eta k_b/c)]\} \tag{28}$$

which can be interpreted as representing a similar disk stability problem without internal damping but with increased air-film damping coefficient and with decreased rotation rate. Thus the effect of introducing η is to decrease the slope of the $v_d{}^2$ line in Fig. 3 which extends the operating range up to a larger critical speed. The intersection point of the two lines in Fig. 3 moves off to infinity when the lines becomne parallel; i.e., the disk is stable for all finite speeds when

$$\frac{\eta k_b}{c} > [1 - v + (3+v)/n]^{-1/2} - 1. \tag{29}$$

The effect of including air-film stiffness ($k_a > 0$) is to introduce another contribution to $v_P{}^2$ which is proportional to Ω^2 which increases the slope of the $v_P{}^2$ line in Fig. 3 and also extends the stable operating range. The air-film stiffness and damping both depend inversely on the film thickness h_a at the hub but these have opposite effects on the critical speed. In the case of small internal damping ($\eta k_b \ll c$), changes in c have little effect on the critical speed and the major effect of decreasing h_a is to increase k_a and thereby extend the stable operating range. It should be mentioned that in the case of a complete disk with no central hole ($a = 0$) the inner boundary condition for pressure becomes $dp/dr = 0$ at $r = 0$ which implies $N[w_r] \equiv 0$ and thus that the air-film stiffness vanishes completely. In the case of a flow seal at the inner rim which introduces the boundary condition $dp/dr = 0$ at $r = a$ the air-film stiffness is considerably reduced in comparison to the case where, with all other factors the same, the flow is unobstructed and the boundary condition is $p = 0$ at $r = a$.

4 Numerical analysis

Finite difference analysis was performed to calculate the precise critical velocity and the mode shape, and to show the validity of the single-mode analysis. In this section internal damping in the disk is neglected.

The matrix formulations of (5) and (6) are given by

$$\lambda^2 \mathbf{M}w + \mathbf{K}_b w + \mathbf{K}_m w = \mathbf{p} \tag{30}$$

$$\mathbf{R}p + \mathbf{N}w = (2\lambda - in\Omega)\,w \tag{31}$$

where M, K_m, R, and N are matrices which represent finite difference approximations of the operators $\varrho_d H$, K_b, K_m, R and N, and w and p are column vectors representing w_r and p_r. Elimination of p from (30) and (31) gives a complex eigenvalue problem.

$$\lambda^2 Mw + (\lambda - in\Omega/2)\,Cw + Kw = 0 \tag{32}$$

where

$$C = R^{-1}, \quad K = K_b + K_m + R^{-1}N. \tag{33}$$

This equation is rewritten as a double-sized eigenvalue equation

$$\begin{bmatrix} 0 & I \\ M^{-1}\left(-K + \dfrac{in\Omega}{2}\,C\right) & -M^{-1}C \end{bmatrix} \begin{bmatrix} w \\ \lambda w \end{bmatrix} = \lambda \begin{bmatrix} w \\ \lambda w \end{bmatrix} \tag{34}$$

which is solved by a standard eigenvalue program.

Eigenvalues and vectors at the critical condition can be obtained by a real eigenvalue analysis which is more easily solved than (34). Since λ is purely imaginary and w is real at the critical condition, equation (32) can be split into two real equations

$$-\omega^2 Mw + Kw = 0 \tag{35}$$

$$\omega = n\Omega/2 \tag{36}$$

where $\lambda = i\omega$. In the actual calculation, first Ω is set equal to an arbitrary value, and ω is then calculated from (35) and checked to see whether (36) is satisfied. If ω is larger than $n\Omega/2$, then Ω is increased, and if not, Ω is decreased. This procedure is repeated until the solution converges.

5 Results

Critical speed for an 8-inch floppy disk calculated by the single-mode analysis, Eq. (21), and by the numerical analysis, Eqs. (35) and (36), are shown in Fig. 4 for the case where disk internal damping is neglected. In one case the airfilm stiffness k_a is taken to have the value (17) while in the other case k_a was set equal to zero. The two methods show good agreement for both cases. They show the characteristics predicted in the previous section; the critical velocity for $k_a = 0$ is

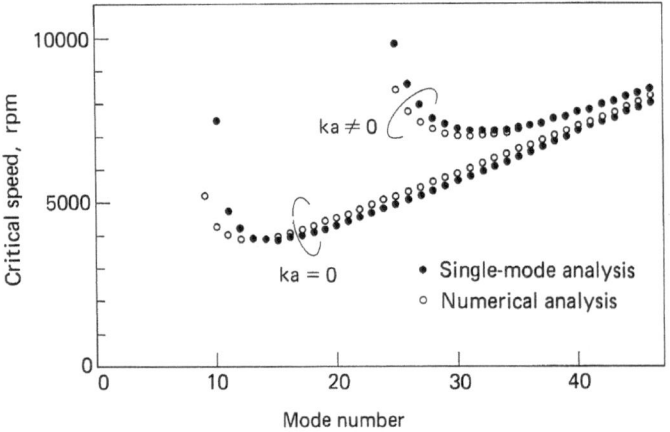

Fig. 4. Critical speeds of 8-inch flexible disk system predicted by single-mode and numerical analyses

smaller than the one for $k_a \neq 0$ and the critical velocities take their minimum value at around $n = 13$ for $k_a = 0$ and $n = 30$ for $k_a \neq 0$. It is also found that the difference between the solutions for $k_a = 0$ and $k_a \neq 0$ becomes smaller as n becomes larger. This happens because k_a is proportional to $1/n$ when n is large, as seen in (17). It is found that the minimum critical speeds with and without air-film stiffness take values of the same order, namely 3 500 rpm and 7 000 rpm. Therefore the minimum velocity by the single mode analysis without air-film stiffness, which is given by (26), is a simple and practical tool for roughly estimating the lower bounds for the critical velocity.

Mode shapes at the critical condition obtained by numerical analysis are shown in Fig. 5. Figure 5a shows disk deflection and Fig. 5b shows air pressure for the 30th mode. They have large values only near the outer edge, which confirms the validity of the assumption in the previous section that ϕ, ψ and their derivatives are only large in the neighborhood of $r = b$.

Figures 6a and 6b show disk deflection for under- and over-critical conditions in another disk, where the steady state deflection w_0 is superposed on the unsteady deflection $Re[w']$. Figures 6c and 6d show normalized deflections $Re[w']/Abs[w']$ so that the amplitude is exaggerated at the inner part. Although there is not a visible difference between the deflections in 6a and 6b, the curvatures of the nodal lines in 6c and 6d have opposite sense. We believe this phenomenon is a consequence of the nonproportional nature of the fluid-film damping. See Appendix for further discussion.

The effect of disk internal damping is shown in Fig. 7 for the 8-inch floppy disk system obtained by the single-mode analysis. The figure shows that η has a stabilizing effect, i.e., the critical speed for $\eta = 0.000\,2$ s is about 40% greater than that for the undamped case and the critical speed for $\eta = 0.000\,4$ s is more than twice that for the undamped case. The instability disappears completely for $\eta > 0.000\,52$ s.

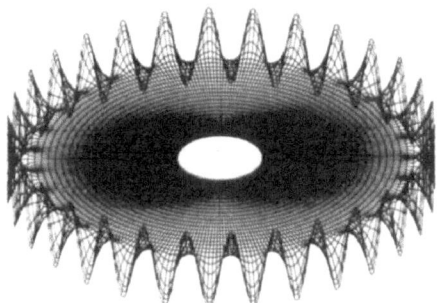

(a) Disk deflection

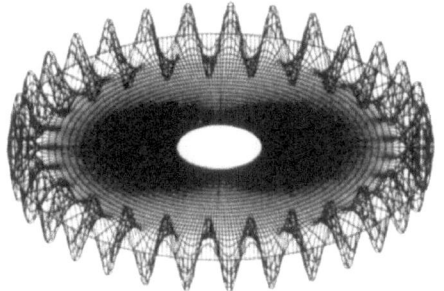

(b) Air pressure   **Fig. 5.** Disk deflection and air film pressure at critical speed

The minimum critical speeds for various types of flexible disks calculated according to (26) are shown in Fig. 8. They give lower bounds of the speed since air-film stiffness and disk internal damping are neglected. For commercial floppy disks, 3.5″ FD, 5.25″ FD and 8″ FD, the critical speeds are higher than 3 600 rpm, which is above the usual disk velocity. Thus, unstable vibrations do not appear in conventional disk drives.

When the disks are made of VCR-foil, the critical speeds are drastically lower since thay are only about one tenth as thick as commercial floppy disks.The 8-inch VCR-foil disk, which has the lowest value of all cases, has a lower bound for the critical speed of 450 rpm. This value is smaller than that of commercial non-contact disk drives. Therefore, if VCR-foils are to be used as practical computer disks, special efforts must be exerted to introduce design features which delay the onset of instability.

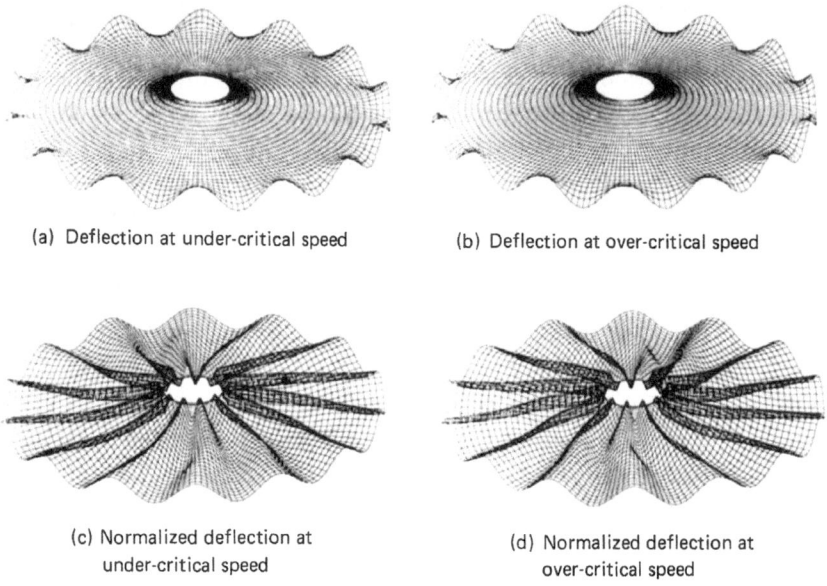

(a) Deflection at under-critical speed

(b) Deflection at over-critical speed

(c) Normalized deflection at
under-critical speed

(d) Normalized deflection at
over-critical speed

Fig. 6. Disk deflection at supercritical and subcritical speeds

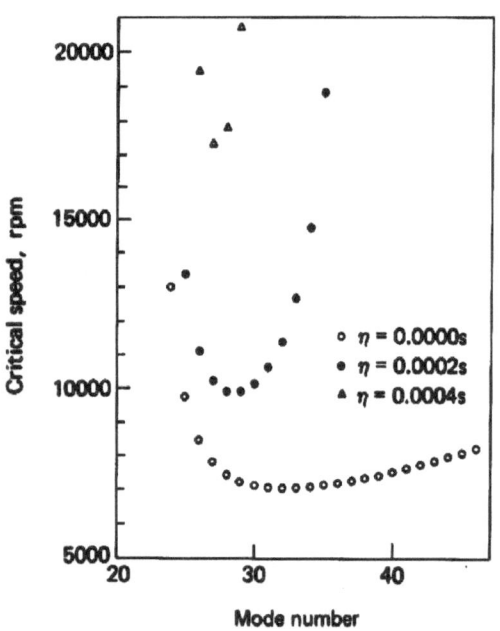

○ $\eta = 0.0000s$
● $\eta = 0.0002s$
▲ $\eta = 0.0004s$

Fig. 7. Effect of internal damping on critical speed

No instability for $\eta > 0.00052s$

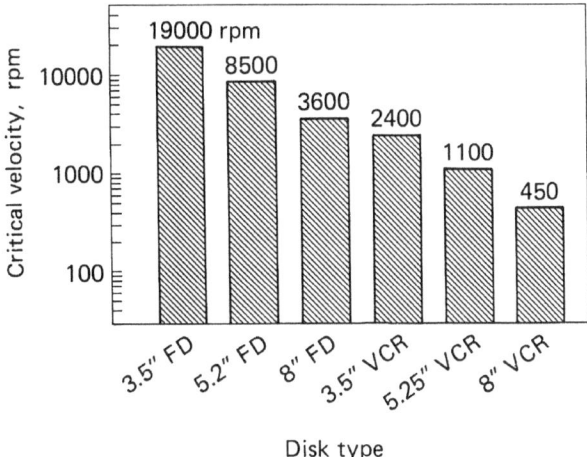

Fig. 8. Lower bounds of critical speed for various types of flexible disk

6 Conclusion

Self-excited vibrations of a flexible disk rotating above a flat surface were studied analytically to provide insights for the design of high-speed magnetic disk drives. First, a single mode analysis was performed to obtain a rough estimate of the critical speed and to clarify its physical meaning. Next, finite difference analysis was performed to calculate the precise critical speed and to show the validity of the single-mode analysis. Thse analyses show that:

— Instability appears when the relative velocity between the disk and the average air flow is larger than the travelling wave velocity of the disk vibration.

— The stability of the system is improved as the disk thickness, Young's modulus, air density and the disk internal damping are increased, and the disk outer radius, disk density and the air film thickness are decreased. Air flow at the disk center also improves stabilities.

— Critical speeds for commercially available floppy disks are higher than 3 600 rpm, whereas a lower bound estimate of the critical velocity for an 8-inch VCR-foil disk without air film stiffness and internal damping is 450 rpm.

Appendix

Curvature of Nodal Lines

The disk displacement (3) represents a travelling wave propagating in the circumferential direction. In general the parameter λ and the function $w_r(r)$ are complex. The imaginary part of λ fixes the wave frequency while the real part determines the temporal growth rate. If the phase of $w_r(r)$ is independent of r the waves will be straight crested; i.e., the nodal lines will be n equally spaced diameters. If the phase varies with r the nodal lines will be curved. At the critical condition described by (35) and (36) w is real and the nodal lines are straight. We present here a perturbation analysis which suggests that the curvature of the nodal lines will have opposite sense for subcritical and supercritical spereds if the damping is non-proportional.

Consider Eq. (32) for the radial mode shape w of the wave with n nodal lines. The inertia matrix M is symmetric and independent of the speed Ω, while the damping and stiffness matrices

C and K, are generally unsymmetric and speed-dependent. Let the fundamental solution (32) at the critical speed Ω_1 be $\lambda = i\omega_1 = in\Omega_1/2$ and $w = w_1$. We consider linearized perturbations in this neighborhood under the assumption that the matrices C and K remain constant. The order N of the matrices depends on the number of radial subdivisions employed in the finite difference approximation. For the eigenvalue problem (35) let the matrix of right eigenvectors w_j, $j = 1, ..., N$, be denoted by Φ and let the matrix of left eigenvectors v_j, $j = 1, ..., N$, be denoted by Ψ. Then the matrix $\Psi^t M \Phi$ is a diagonal matrix of modal masses m_j. We take the eigenvectors to be so normalized that the modal masses are all equal; i.e., $m_j = m$, $j = 1, ..., N$. The matrix $\Psi^t K \Phi$ is a diagonal matrix of modal stiffnesses $k_j = m\omega_j^2$. In general the matrix $\Psi^t C \Phi$ is a full matrix $\bar{C}$ with elements $\bar{c}_{ij}$ which have no symmetry properties. If the damping were proportional then $\Psi^t C \Phi$ would be a diagonal matrix of modal damping coefficients c_j.

The perturbation analysis of (32) is carried out by setting $\Omega = \Omega_1(1 + \varepsilon)$, where ε is a small perturbation parameter, and retaining only first-order changes in λ and w. We find

$$\lambda = i\omega_1 \left(1 + \varepsilon \frac{\bar{c}_{11}}{\bar{c}_{11} + 2i\omega_1 m} \right) \tag{37}$$

and

$$w = \left(1 - \varepsilon \frac{2}{\bar{c}_{11} + 2i\omega_1 m} \Phi D \Psi^t C \right) w_1 \tag{38}$$

where D is the diagonal matrix whose elements are

$$d_{11} = 0, \quad d_{jj} = \frac{\omega_1^2}{\omega_j^2 - \omega_1^2}, \quad j = 2, ..., N. \tag{39}$$

An alternative representation for w is

$$w = w_1 - \varepsilon \sum_{j=2}^{N} \frac{2\bar{c}_{j1}}{\bar{c}_{11} + 2i\omega_1 m} \frac{\omega_1^2}{\omega_j^2 - \omega_1^2} w_j. \tag{40}$$

With $\bar{c}_{11}$ real and positive (37) indicates that as the speed Ω passes from subcritical to supercritical the wave frequency increases monotonically while the temporal behaviour changes continuously from decay at subcritical speeds to growth at supercritical speeds. This behavior is independent of whether the damping is proportional or not. From (40) it follows that there is no first-order change in the eigenvector with speed for proportional damping ($\bar{c}_{ij} = 0$, $i \neq j$) which implies that the nodal lines would remain straight. In the case of non-proportional damping (40) shows that the phase of the eigenvector varies with position r, which introduces curvature into the nodal lines. Furthermore the sense of the curvature is reverse when the sign of ε is changed. Thus the curvature of the nodal lines for supercritical (growing) waves is opposite from the curvature of the nodal lines for subcritical (decaying) waves.

Acknowledgement

The authors are grateful to Kay Herbert of M.I.T. for his valuable help in the development of the numerical analysis program.

References

Adams, G. G.: Analysis of the flexible disk/head interface. J. Lubr. Technol. **102**/1, 86−90 (1980).

Adams, G. G.: Critical speeds for a flexible spinning disk. Int. J. Mech. Sci. **29**/8, 525−531 (1987).

Bilharz, H.: Bemerkungen zu einem Satz von Hurwitz. ZAMM **24**, 77−82 (1944).

Bogy, D. G., Talke, F. E.: Steady axisymmetric solutions for pressurized rotating flexible disk packs. IBM J. Res. Develop. **22**/2, 179−184 (1978).

Hosaka, H., Nishida, Y.: An elastohydrodynamic analysis of back-plate-type foil disks. ASLE SP-22, 167−173 (1987).

Itao, K., Crandall, S. H.: Natural modes and natural frequencies of uniform, circular, free-edge plates. J. Appl. Mech. **46**/2, 448−453 (1979).

Lamb, H., Southwell, R.: The vibrations of a spinning a disk. Proc. Roy. Soc. (London) A **99**, 272−280 (1921).

Licari, J. P., King, F. K.: Elastohydrodynamic analysis of head to flexible disk interface phenomena. J. Appl. Mech. **48**/4, 763−768 (1981).

Ono, K., Maeno, T.: Theoretical and experimental investigation on dynamic characteristics of a 3.5-inch flexible disk due to a point contact head. ASLE SP-21, 144−151, 1986.

Pearson, R. T.: The development of the flexible-disk magnetic recorder. Proc. IRE **49**/1, 164−174 (1961).

Pelech, I., Shapiro, A. H.: Flexible disk rotating on a gas film next to a wall. J. Appl. Mech. **31**/4, 577−584 (1964).

Pinkus, O., Lund, J. W.: Centrifugal effects in thrust bearings and seals under laminar conditions. J. Lubr. Technol. **103**/1, 126−136 (1981).

Robertson, D.: Whirling of a journal in a sleeve bearing. Phil. Mag. [7] **15**, 113−130 (1933).

Sato, Y.: Dynamic behavior of a very flexible membrane rotating on a gas film next to a wall. Acta Mechanica **55**, 95−104 (1985).

Smith, D. M.: The motion of a rotor carried by a flexible shaft in flexible bearings. Proc. Roy. Soc. (London) A **142**, 92−118 (1933).

Authors' address: Prof. S. H. Crandall, Department of Mechanical Engineering, Massachusetts Institute of Technology, Cambridge, MA 02139, U.S.A.

Acta Mechanica (1992) [Suppl] 3: 129–145

Bifurcation of nonlinear normal modes
in a class of two degree of freedom systems

R. H. Rand, New York, U.S.A., **C. H. Pak,** Inchon, Korea, and **A. F. Vakakis,** Illinois, U.S.A.

Summary. This work concerns the nonlinear vibrations of a class of two-degree-of-freedom autonomous conservative systems consisting of two coupled nonlinear oscillators with cubic coupling forces. For such a system, nonlinear normal modes (NNM's) have long been studied as "vibrations in unison", i.e. periodic motions in which x and y simultaneously achieve zero velocity. We investigate the stability and bifurcation of NNM's by using a perturbation method together with the computer algebra system MACSYMA.

The results of our study include the characterization of those systems which are maximally degenerate in the sense of changing the number of elementary periodic motions which they exhibit in response to a small change in parameters. By looking in the neighborhood of such a degenerate system, we obtain universal unfoldings. In particular we show how NNM's, which project onto the configuration plane as (curved) line segments, are related to a family of periodic orbits which look like ellipses on the configuration plane. Our analytic results are shown to compare favorably with the results of numerical integration.

1 Introduction

This work concerns a class of periodic motions (PM's) called nonlinear normal modes (NNM's). NNM's may be thought of as nonlinear continuations of the linear normal modes (LNM's) associated with eigenvalue problems in a linearized system. Interest in LNM's may be said to stem both from their role as basis vectors for the general solution of a linear system, as well as from their role as sources of large deflection of forced linear structures due to the phenomenon of resonance. Although the principle of superposition cannot be applied to nonlinear systems, early interest in NNM's focused on nonlinear resonance. NNM's were studied by Kauderer [5], Rosenberg [16] and others [4, 9, 10, 12, 13] since the 1960's. More recently, Vakakis [19] examined NNM's from the point of view of modern dynamics. It was shown [16] that NNM's can be either "similar" or "nonsimilar". Similar modes correspond to straight lines in the configuration space of the system (and in this sense they resemble LNM's), whereas nonsimilar NNM's correspond to curved lines. Methods for detecting NNM's in discrete nonlinear systems already exist in the literature [8, 14, 16]. Recently invariant manifold concepts have been utilized for this purpose [17].

In previous works [7, 9] it was shown that nonlinear forced resonance occurs in the neighborhood of NNM's. Moreover, it was shown that the bifurcations of NNM's affect the topological structure of the resonance curves in a class of systems [19], and their global dynamics [20]. The study of NNM's and their bifurcations is thus of practical importance for an understanding of the dynamics of nonlinear oscillators.

Until now, only bifurcations of NNM's to other NNM's were studied in the literature. The present work views NNM's as they are related to more general PM's. According to "Poincare's conjecture" ([18], p. 381) phase space in a compact Hamiltonian system is dense with PM's. In some nonlinear systems [2] PM's have been categorized by their topology into families which are connected by bifurcations due to changes in parameters. These branches of PM's are similar to branches of equilibria encounted in nonlinear buckling problems [1]. Our goal is to investigate how NNM's bifurcate, i.e. how they change their number and form in response to a change in parameters. We conduct our study in a general class of two degree of freedom systems involving six independent parameters. We use the two variable expansion perturbation method and we use computer algebra (MACSYMA) to handle the large quantities of algebra involved in the computations. Since the perturbation method is valid in the small amplitude limit, our results are necessarily approximate. To within this approximation, we obtain a closed form solution in the form of a first integral for the slow flow. Using this result, we determine the most degenerate singularity involving a confluence of PM's, and we obtain a universal unfolding of this singularity, thereby revealing how the nearby families of PM's are related.

2 Perturbation method

We consider the general class of two degree of freedom conservative autonomous systems with kinetic energy T:

$$T = \frac{1}{2}\left[\left(\frac{dx}{dt}\right)^2 + \left(\frac{dy}{dt}\right)^2\right] \tag{1}$$

and potential energy V:

$$V = V_1 + \varepsilon V_2 \tag{2}$$

where

$$V_1 = \frac{1}{2}(x^2 + \omega^2 y^2), \tag{3}$$

$$V_2 = a_{40}x^4 + a_{31}x^3y + a_{22}x^2y^2 + a_{13}xy^3 + a_{04}y^4 \tag{4}$$

and where

$$\omega^2 = 1 + \varepsilon\Delta. \tag{5}$$

This system evolves according to the following two differential equations:

$$\frac{d^2x}{dt^2} + x = -\varepsilon\frac{\partial V_2}{\partial x}, \tag{6}$$

$$\frac{d^2y}{dt^2} + (1 + \varepsilon\Delta)\,y = -\varepsilon\frac{\partial V_2}{\partial y}, \tag{7}$$

and is characterized by the five nonlinear restoring force coefficients a_{ij} and a detuning coefficient Δ. The parameter ε is included in order to provide a scaling for the perturbation method, and so we assume $\varepsilon \ll 1$ and we neglect terms of order ε^2 throughout. Note that when $\varepsilon = 0$ the system is in 1:1 resonance so that we are perturbing off of a system in which the linear normal modes are degenerate.

In order to obtain an approximate solution for the system (1)−(7), we use the two-variable expansion method [6, 11, 15]. We replace t as independent variable by the two variables

$$\xi = t \quad \text{and} \quad \eta = \varepsilon t \tag{8}$$

in which case Eq. (6) becomes

$$\frac{\partial^2 x}{\partial \xi^2} + 2\varepsilon \frac{\partial^2 x}{\partial \xi\, \partial \eta} + \varepsilon^2 \frac{\partial^2 x}{\partial \eta^2} + x = -\varepsilon \frac{\partial V_2}{\partial x} \tag{9}$$

and where Eq. (7) is similarly transformed. Next we expand x and y in power series in ε,

$$x(\xi, \eta, \varepsilon) = x_0(\xi, \eta) + \varepsilon x_1(\xi, \eta) + \cdots \tag{10}$$

and a similar expression for y. Substitution into the d.e.'s on x and y (e.g. Eq. (9)) gives, after collecting terms and equating the coefficients of like powers of ε to zero:

$$\frac{\partial^2 x_0}{\partial \xi^2} + x_0 = 0, \quad \frac{\partial^2 y_0}{\partial \xi^2} + y_0 = 0, \tag{11}$$

$$\frac{\partial^2 x_1}{\partial \xi^2} + x_1 = -2 \frac{\partial^2 x_0}{\partial \xi\, \partial \eta} - \frac{\partial V_2}{\partial x}\bigg|_{x=x_0, y=y_0} \tag{12}$$

and a similar eq. on y_1. We take the solution of Eqs. (11) in the form:

$$x_0 = A(\eta) \cos \xi + B(\eta) \sin \xi, \tag{13}$$

$$y_0 = C(\eta) \cos \xi + D(\eta) \sin \xi \tag{14}$$

where $A(\eta), B(\eta), C(\eta), D(\eta)$ are arbitrary functions of η. We substitute Eqs. (13), (14) into the d.e.'s on x_1 and y_1 (e.g. Eq. (12)), trigonometrically simplify and collect terms, and equate to zero the coefficients of the resonant terms $\cos \xi$ and $\sin \xi$, giving:

$$\frac{dA}{d\eta} = \frac{1}{8} \left[3a_{13}(C^2 + D^2)\, D + 3a_{31}(A^2 + 3B^2)\, D + 2a_{22}(C^2 + 3D^2)\, B \right.$$

$$\left. + (4a_{22}D + 6a_{31}B)\, AC + 12a_{40}(A^2 + B^2)\, B \right] \tag{15}$$

and three similar eqs. on $B(\eta)$, $C(\eta)$, $D(\eta)$. Next we transform to polar coordinates, $R_1(\eta)$, $R_2(\eta)$, $\theta_1(\eta)$, $\theta_2(\eta)$

$$A = R_1 \cos \theta_1, \quad B = R_1 \sin \theta_1, \quad C = R_2 \cos \theta_2, \quad D = R_2 \sin \theta_2 \tag{16}$$

noting that the zeroth order solution, Eqs. (13), (14), can be written

$$x_0 = R_1(\eta) \cos \left(\xi - \theta_1(\eta)\right), \quad y_0 = R_2(\eta) \cos \left(\xi - \theta_2(\eta)\right). \tag{17}$$

Substituting Eqs. (16) into the d.e.'s on A, B, C, D (e.g. Eq. (15)) we obtain d.e.'s on R_1, R_2, θ_1, θ_2 which are naturally written in terms of the variable φ:

$$\varphi = \theta_2 - \theta_1, \tag{18}$$

$$\frac{dR_1}{d\eta} = \frac{1}{4} a_{22} R_1 R_2{}^2 \sin 2\varphi + \frac{3}{8} [a_{13} R_2{}^3 + a_{31} R_1{}^2 R_2] \sin \varphi, \tag{19}$$

$$\frac{dR_2}{d\eta} = -\frac{1}{4} a_{22} R_1{}^2 R_2 \sin 2\varphi - \frac{3}{8} [a_{13} R_1 R_2{}^2 + a_{31} R_1{}^3] \sin \varphi, \tag{20}$$

$$\frac{d\theta_1}{d\eta} = -\frac{1}{4} a_{22} R_2{}^2 [\cos 2\varphi + 2] - \frac{3}{8} a_{13} \frac{R_2{}^3}{R_1} \cos \varphi$$

$$\qquad\qquad - \frac{9}{8} a_{31} R_1 R_2 \cos \varphi - \frac{3}{2} a_{40} R_1{}^2, \tag{21}$$

$$\frac{d\theta_2}{d\eta} = -\frac{\varDelta}{2} - \frac{1}{4} a_{22} R_1{}^2 [\cos 2\varphi + 2] - \frac{3}{8} a_{31} \frac{R_1{}^3}{R_2} \cos \varphi$$

$$\qquad\qquad - \frac{9}{8} a_{13} R_1 R_2 \cos \varphi - \frac{3}{2} a_{04} R_2{}^2. \tag{22}$$

Subtracting Eq. (21), (22) and using Eq. (18), we obtain a differential equation on $\varphi(\eta)$:

$$\frac{d\varphi}{d\eta} = -\frac{\varDelta}{2} + \frac{3}{8} a_{13} \cos \varphi \left[\frac{R_2{}^3}{R_1} - 3 R_1 R_2\right]$$

$$\qquad\qquad - \frac{3}{8} a_{31} \cos \varphi \left[\frac{R_1{}^3}{R_2} - 3 R_1 R_2\right] - \frac{3}{2} [a_{04} R_2{}^2 - a_{40} R_1{}^2]$$

$$\qquad\qquad + \frac{1}{4} a_{22} [R_2{}^2 - R_1{}^2] (\cos 2\varphi + 2). \tag{23}$$

Finally we transform to polars R, ψ in the $R_1 - R_2$ plane,

$$R_1 = R \cos \psi, \quad R_2 = R \sin \psi \tag{24}$$

which replaces Eqs. (19), (20), (23) by the following:

$$\frac{dR}{d\eta} = 0, \tag{25}$$

$$\frac{d\psi}{d\eta} = -\frac{1}{8} R^2 a_{22} \sin 2\varphi \sin 2\psi + \frac{3}{16} R^2 \sin \varphi [(a_{13} - a_{31}) \cos 2\psi - (a_{13} + a_{31})], \tag{26}$$

$$\frac{d\varphi}{d\eta} = -\frac{\Delta}{2} + R^2 \left(-\frac{3}{8} a_{31} \cos \varphi [\cos^2 \psi \cot \psi - \frac{3}{2} \sin 2\psi] \right.$$

$$+ \frac{3}{8} a_{13} \cos \varphi \left[\sin^2 \psi \tan \psi - \frac{3}{2} \sin 2\psi \right]$$

$$\left. - \frac{3}{2} [a_{04} \sin^2 \psi - a_{40} \cos^2 \psi] - \frac{1}{4} a_{22} \cos 2\psi (\cos 2\varphi + 2) \right). \tag{27}$$

From Eq. (25), we see that R is a constant, a result equivalent to conservation of energy. From Eqs. (26), (27) we may obtain the single first order differential equation:

$$\frac{d\varphi}{d\psi} = \frac{f(\varphi, \psi)}{g(\varphi, \psi)} \tag{28}$$

where

$$f(\varphi, \psi) = \left[4\frac{\Delta}{R^2} - 2a_{22} + 12a_{04} \right] \cos \psi \sin \psi$$

$$+ [4a_{22} - 12(a_{40} + a_{04})] \cos^3 \psi \sin \psi$$

$$+ [-3a_{13} + (15a_{13} - 9a_{31}) \cos^2 \psi + 12(a_{31} - a_{13}) \cos^4 \psi] \cos \varphi$$

$$+ 4a_{22} \cos^2 \varphi \sin \psi \cos \psi \cos 2\psi, \tag{29}$$

$$g(\varphi, \psi) = [3(a_{31} - a_{13}) \cos^2 \psi + 3a_{13} + 4a_{22} \cos \varphi \cos \psi \sin \psi] \sin \varphi \cos \psi \sin \psi. \tag{30}$$

In order to integrate Eq. (28), we write it in the form $g \, d\varphi - f \, d\psi = 0$, which turns out to be an exact differential. This leads us to the following first integral of Eqs. (28)−(30):

$$-\frac{\Delta}{R^2} \cos 2\psi + \frac{3}{8} (a_{31} - a_{13}) \cos \varphi \sin 4\psi + \frac{3}{4} (a_{31} + a_{13}) \cos \varphi \sin 2\psi$$

$$+ \frac{3}{8} (a_{40} + a_{04}) \cos 4\psi + \frac{3}{2} (a_{40} - a_{04}) \cos 2\psi + \frac{1}{4} a_{22} \cos^2 \varphi$$

$$- \frac{1}{8} a_{22} [2 + \cos 2\varphi] \cos 4\psi = \text{constant}. \tag{31}$$

For a given set of parameters a_{ij} and Δ/R^2, we may plot the integral curves of Eq. (31) on the $\psi - \varphi$ phase plane, for $0 < \psi < \pi/2$, $0 \le \varphi \le 2\pi$. We exclude the line $\psi = 0$ since on this line R_2 vanishes and θ_2 is not defined. Similarly we exclude the line $\psi = \pi/2$ since on this line R_1 vanishes and θ_1 is not defined. In fact, the motion $R_2 \equiv 0$ corresponds to the "x-mode", $y \equiv 0$, which exists iff $a_{31} = 0$. Similarly, the motion $R_1 \equiv 0$ corresponds to the "y-mode", $x \equiv 0$, and it

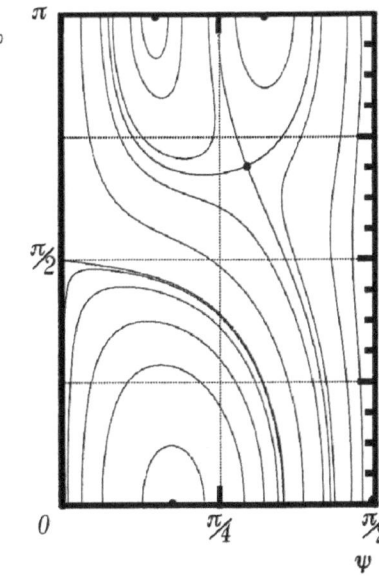

Fig. 1. Approximate first integral, Eq. (31), for parameters (32)

exists iff $a_{13} = 0$. Thus these special PM's (which are "similar" NNM's) may be investigated directly from Eqs. (6), (7) and will not be considered in our perturbation result Eq. (31).

Note that the first integral (31) is invariant under the transformation $\varphi \to 2\pi - \varphi$, and thus the phase portrait in the region $\pi \leqq \varphi \leqq 2\pi$ is, when reflected about the line $\varphi = \pi$, the mirror image of that in $0 \leqq \varphi \leqq \pi$. Therefore we may display our results in the region $0 < \psi < \pi/2$, $0 \leqq \varphi \leqq \pi$. See Fig. 1, where we display Eq. (31) for the parameter values:

$$a_{13} = 0, \quad a_{22} = 1, \quad a_{31} = 1, \quad a_{04} = 0, \quad a_{40} = \frac{1}{4}, \quad \Delta/R^2 = 0. \tag{32}$$

In the next section we will discuss the significance of some of the details in Fig. 1.

3 Periodic motions

Examination of Fig. 1 shows the presence of singular points. Such points represent solutions at which φ and ψ are fixed in time, which, from (24), (21), (22) and (17) mean that the amplitudes R_1 and R_2 are fixed, the phases θ_1 and θ_2 are linear in slow time η, and the phase difference is fixed between the x and y motions. Thus to $0(\varepsilon)$ such motions represent PM's of the system (1)−(7). In the case that φ is 0 or π, the x and y motions represent vibrations-in-unison, both motions simultaneously achieving their maximum displacements and velocities. These PM's plot as a single-valued cue $y = y(x)$ in the $x - y$ configuration plane (approximately a straight line for small ε). Such motions were defined by Rosenberg [16] as NNM's.

In contrast to NNM's, a singular point in Fig. 1 which corresponds to a value of φ which is unequal to 0 or π represents a PM which is not a NNM. To investigate the form of such PM's, we rewrite Eq. (17.2) using Eq. (18):

$$y_0 = R_2 \cos(\xi - \theta_1 - \varphi) = R_2[\cos(\xi - \theta_1)\cos\varphi + \sin(\xi - \theta_1)\sin\varphi] \tag{33}$$

which becomes, using Eq. (17.1),

$$y_0 = R_2 \left[\frac{x_0}{R_1} \cos\varphi + \sqrt{1 - \left(\frac{x_0}{R_1}\right)^2} \sin\varphi \right]. \tag{34}$$

Eq. (34) may be simplified so that it takes the form of a conic section:

$$[x_0 \quad y_0] \begin{bmatrix} k_{11} & k_{12} \\ k_{21} & k_{22} \end{bmatrix} \begin{bmatrix} x_0 \\ y_0 \end{bmatrix} = 1 \tag{35}$$

where

$$k_{11} = (R_1 \sin \varphi)^{-2},$$

$$k_{12} = k_{21} = -(R_1 R_2 \tan \varphi \sin \varphi)^{-1}, \tag{36}$$

$$k_{22} = (R_2 \sin \varphi)^{-2}.$$

The trace and determinant of the matrix $[k_{ij}]$ are positive for such PM's:

$$\text{trace} = \left(\frac{1}{R_1{}^2} + \frac{1}{R_2{}^2} \right) \frac{1}{\sin^2 \varphi} > 0, \tag{37}$$

$$\text{determinant} = [R_1 R_2 \sin \varphi]^{-2} > 0. \tag{38}$$

Thus the eigenvalues of $[k_{ij}]$ are real and positive, and the orbit of such a PM is an ellipse on the $x - y$ plane. We shall denote such a PM as an elliptic orbit (EO). EO's correspond to singularities in the $\varphi - \psi$ plane for which $\varphi \neq 0, \pi$, cf. Fig. 1.

Note that although both $\psi = 0$ and $\psi = \pi/2$ are integral curves in Fig. 1, they respectively correspond to the vanishing of R_2 and R_1, at which points θ_2 and θ_1 are respectively undefined. In order to investigate the existence of motions corresponding to the vanishing of R_2 or R_1, we proceed directly from Eqs. (6), (7). These show that since $a_{13} = 0$, the y-mode, $x \equiv 0$ is a PM. However, since $a_{31} = 1$, the x-mode, $y \equiv 0$, is not a PM of the system. We conclude that on the lines $\psi = 0, \pi/2$ the preceding analysis gives spurious results and is inapplicable. In particular the singular point found on the line $\psi = 0$ in Fig. 1 must not be interpreted as a PM.

In addition to revealing the existence of PM's, the first integral (31) also discloses their stability. The stability of the PM in the original system (6), (7) is the same as the stability of the singular point in the slow flow (26), (27). Since the system is conservative, only centers and saddles can generically occur, and so their stability is easily discerned. E.g. in Fig. 1, the three NNM's are stable while the EO is unstable.

An approximation for the frequency p of a PM may be obtained from Eq. (17) as follows: First we find the associated values of φ and ψ corresponding to a singularity of Eqs. (26), (27). Then for a given value of R, Eq. (24) yields R_1 and R_2. Substitution of these values into Eqs. (21), (22) gives (equal) values for $\dfrac{d\theta_1}{d\eta} = \dfrac{d\theta_2}{d\eta} = k$ (say) which may be integrated to give

$$\theta_1 = k\eta + \theta_{10} \quad \text{and} \quad \theta_2 = k\eta + \theta_{20}. \tag{39}$$

Then from Eqs. (17) we find the frequency p as:

$$p = 1 + \varepsilon k. \tag{40}$$

4 Numerical simulations

In order to verify the analytical results of the perturbation analysis, we numerically integrated the nonlinear equations of motion (6), (7), using a fourth-order Runge-Kutta algorithm. First, the system with parameters of Eq. (32) was examined, with $\varepsilon = 0.1$ and $R = 1$, cf. Fig. 1. The equations of motion were integrated with initial conditions corresponding to the analytically predicted periodic motions. Figure 2 shows the result for the EO of Fig. 1 which corresponds to $\varphi = 2.17$ and $\psi = 0.923$. The orbital instability of this motion is verified by the projection on the configuration plane of Fig. 2a, and the time response of $x(t)$ in Fig. 2b. All the analytically predicted NNM's were verified, and their orbital stability was confirmed. As an example, Fig. 3 shows the NNM at $\varphi = 0$, $\psi = 0.54$. In the configuration plane of Fig. 3a the motion is approximately a straight line passing through the origin (in accordance with theory), and the time response of $x(t)$ in Fig. 3b contains only small amplitude modulation.

As an example of a system with a stable EO, the following parameters were chosen:

$$a_{13} = 1, \quad a_{22} = 1.4, \quad a_{31} = 0.8, \quad a_{04} = 2.2, \quad a_{40} = 0.5, \quad \Delta = -0.1, \quad R = 0.2. \quad (41)$$

The perturbation result (31) is displayed in Fig. 4. The EO at $\varphi = 2.88$, $\psi = 0.799$ is examined in Fig. 5 for $\varepsilon = 0.1$, verifying its orbital stability. Note the shape of the EO in the configuration

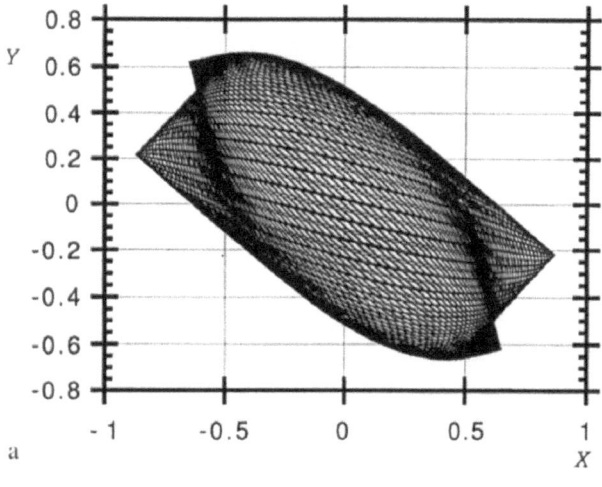

a

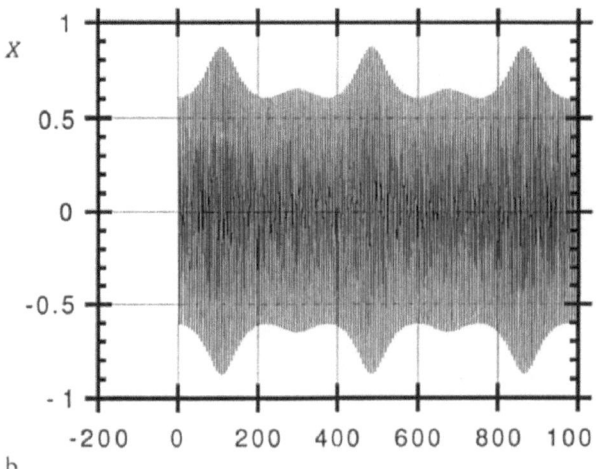

b

Fig. 2. Numerical integration of Eqs. (6), (7) for parameters (32) and $\varepsilon = 0.1$, with initial conditions corresponding to the unstable EO in Fig. 1, $R = 1$, $\varphi = 2.17$ and $\psi = 0.923$

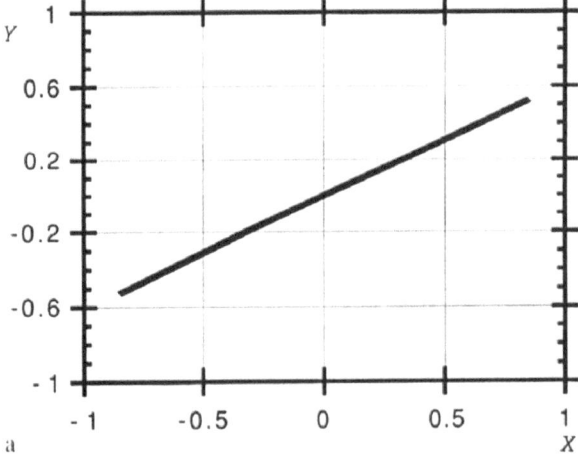

a

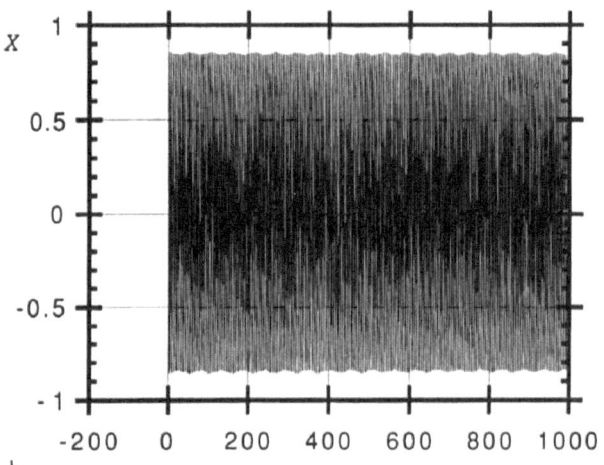

b

Fig. 3. Numerical integration of Eqs. (6), (7) for parameters (32) and $\varepsilon = 0.1$, with initial conditions corresponding to a stable NNM in Fig. 1, $R = 1$, $\varphi = 0$ and $\psi = 0.54$

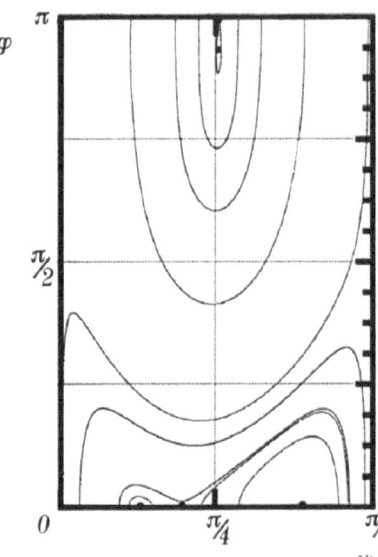

Fig. 4. Approximate first integral, Eq. (31), for parameters (41)

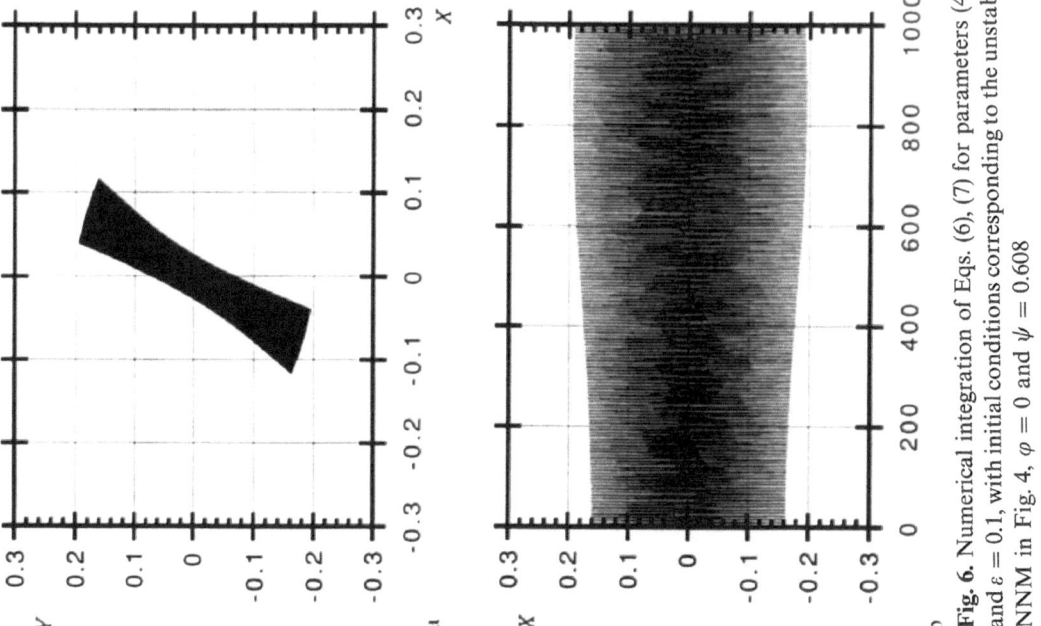

Fig. 6. Numerical integration of Eqs. (6), (7) for parameters (41) and $\varepsilon = 0.1$, with initial conditions corresponding to the unstable NNM in Fig. 4, $\varphi = 0$ and $\psi = 0.608$

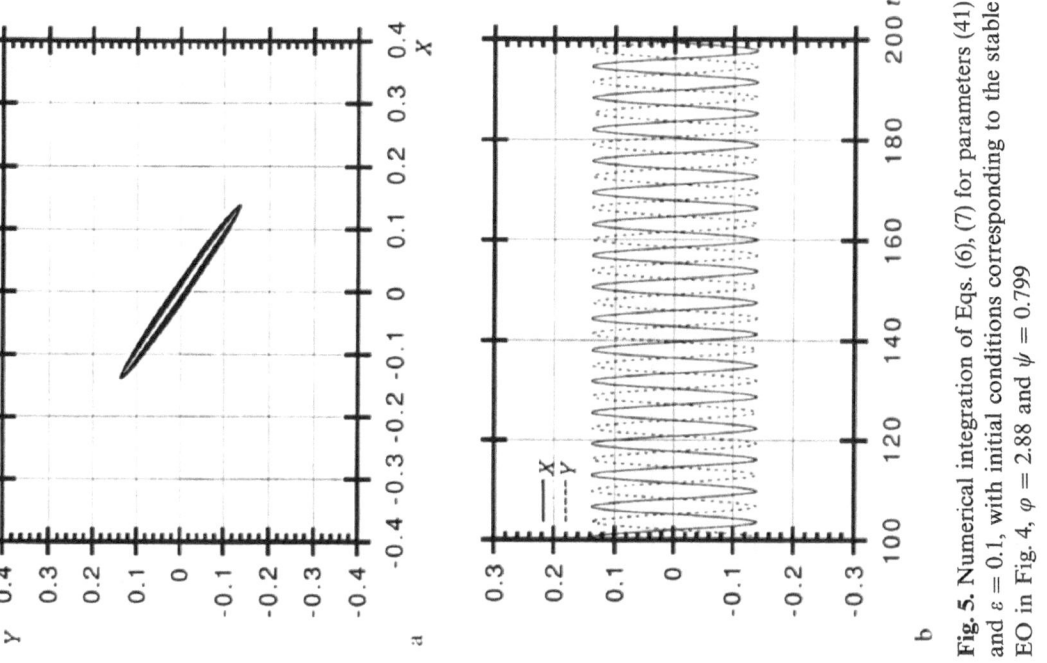

Fig. 5. Numerical integration of Eqs. (6), (7) for parameters (41) and $\varepsilon = 0.1$, with initial conditions corresponding to the stable EO in Fig. 4, $\varphi = 2.88$ and $\psi = 0.799$

plane in Fig. 5a and the phase differences between the coordinates $x(t)$ and $y(t)$ in Fig. 5b. The NNM at $\varphi = 0$, $\psi = 0.608$ is examined in Fig. 6 and is shown to be orbitally unstable, in agreement with the analytical results. Note in this case the large amplitude modulations in the response, an indication of orbital instability.

5 Bifurcations

In the remainder of this paper we shall be interested in the number and type of PM's which occur in our system as a function of the parameter values.

As we move from point to point in the six dimensional parameter space of our system, the singular points of Eq. (28) (which correspond to PM's in the original system) and of its first integral (31) change their position and in exceptional cases change their number and type by merging with other singular points. Although the complete characterization of all such bifurcations in a six dimensional parameter space is too complicated a task, we are interested in characterizing the "worst case", i.e. the most degenerate singularity which may occur. This section contains an analysis of this problem. In the next section we seek a universal unfolding of the degenerate singularity.

The singular points of Eq. (28) are given by simultaneously requiring Eqs. (29) and (30) to be satisfied:

$$f(\varphi, \psi) = 0, \quad g(\varphi, \psi) = 0. \tag{42}$$

From Eq. (30), $g(\varphi, \psi) = 0$ may be satisfied either by taking

$$\sin \varphi = 0 \tag{43}$$

or by taking

$$\cos \varphi = \frac{3(a_{13} - a_{31}) \cos^2 \psi - 3a_{13}}{4a_{22} \cos \psi \sin \psi}. \tag{44}$$

When Eq. (43) is substituted into Eq. (29) for $f(\varphi, \psi) = 0$, we obtain conditions for NNM's (since these occur for $\varphi = 0$ or π). Similarly when Eq. (44) is substituted into $f(\varphi, \psi) = 0$, we obtain conditions for EO's. We will investigate these two cases separately, and we seek conditions on the parameters of the system such that the maximal degeneracy occurs in each. By simultaneously requiring both degeneracy conditions to hold, we will achieve the most degenerate singularity involving the bifurcation of both NNM's and EO's.

We begin by satisfying Eq. (41) with $\varphi = 0$ ($\varphi = \pi$ leads to the same results), which when substituted into Eq. (29) gives the following condition for NNM's:

$$-\frac{4}{3} \frac{\Delta}{R^2} p(p^2 + 1) + a_{13} p^4 + (2a_{22} - 4a_{04}) p^3 + (3a_{31} - 3a_{13}) p^2$$

$$+ (4a_{40} - 2a_{22}) p - a_{31} = 0 \tag{45}$$

where

$$p = \tan \psi.$$

Since Eq. (45) is a quartic on p, the most degenerate case would correspond to a quadruple root, if the parameters could be chosen so that this were possible. This cannot happen however, as may be shown as follows. Let p_0 be the quadruple root. Then

$$(p - p_0)^4 = 0. \tag{46}$$

Comparing coefficients of like powers of p in Eq. (46) and Eq. (45)/a_{13}, we obtain for the powers of p^0 and p^2, respectively:

$$p_0{}^4 = -\frac{a_{31}}{a_{13}}, \qquad 6p_0{}^2 = -3 + 3\frac{a_{31}}{a_{13}}. \tag{47}$$

Eliminating a_{31}/a_{13} in Eq. (47) shows that p_0 must satisfy the eq.

$$p_0{}^4 + 2p_0 + 1 = 0 \tag{48}$$

which has no real roots.

Since a quadruple root is impossible, we next consider that the maximal degeneracy in Eq. (45) could correspond to a triple root, which turns out to be possible. Let p_0 be the triple root, and let p_1 be the remaining root. Then

$$(p - p_0)^3 (p - p_1) = 0. \tag{49}$$

Comparing coefficients of like powers of p in Eq. (49) and Eq. (45)/a_{13}, we obtain four equations from which we can eliminate p_0 and p_1, giving two conditions on the parameters of the problem. Let us call these two resulting conditions Eq. (I) and Eq. (II). Inspection shows that Eq. (I) can be factored into what we shall call Eq. (Ia) and Eq. (Ib):

$$16\left[\frac{\Delta}{R^2}\right]^2 - 48(a_{22} - 2a_{04})\frac{\Delta}{R^2} + 9a_{31}^2 - 54a_{13}a_{31}$$

$$+ 36a_{22}^2 - 144a_{04}a_{22} + 81a_{13}^2 + 144a_{04}^2 = 0, \tag{Ia}$$

$$16a_{31}\left[\frac{\Delta}{R^2}\right]^2 - 48a_{31}(a_{22} - 2a_{04})\frac{\Delta}{R^2} - 9a_{13}^2 + 54a_{13}^2a_{31}$$

$$+ 36a_{22}^2a_{31} - 144a_{31}a_{04}a_{22} - 81a_{13}a_{31}^2 + 144a_{04}^2a_{31} = 0, \tag{Ib}$$

$$314928a_{13}^2a_{22}^2a_{40}\frac{\Delta}{R^2} + 117 \text{ similar terms} = 0. \tag{II}$$

Equation (II) is a homogeneous polynomial of degree 6 in the parameters Δ/R^2, $a_{40}, a_{31}, a_{22}, a_{13}$, a_{04} having 118 terms and is therefore too long to give here. Our conclusion so far is that in order for a system to exhibit a triply degenerate NNM, its parameters must satisfy either Eq. (Ia) or (Ib), and Eq. (II).

We continue by looking for the comparable maximally degenerate EO by satisfying $g(\varphi, \psi) = 0$ with Eq. (44). Substitution of Eq. (44) into Eq. (29) for $f(\varphi, \psi) = 0$ and solving for $\cos^2 \psi$ gives:

$$\cos^2 \psi = \frac{8a_{22}(\Delta/R^2) + 9a_{13}a_{31} - 4a_{22}^2 + 24a_{04}a_{22} - 9a_{13}^2}{-9(a_{13} - a_{31})^2 - 8a_{22}(a_{22} - 3a_{04} - 3a_{40})} \equiv Z. \tag{50}$$

If we represent the expression for $\cos^2 \psi$ in Eq. (50) by Z, Eq. (44) becomes

$$\cos^2 \varphi = \frac{9[(a_{31} - a_{13})\,Z + a_{13}]^2}{16a_{22}^2(1 - Z)\,Z}. \tag{51}$$

Since $\cos^2 \varphi \leqq 1$, the most degenerate case of Eq. (51) corresponds to $\cos^2 \varphi = 1$, which gives a quadratic on Z:

$$[9(a_{31} - a_{13})^2 + 16a_{22}^2]\,Z^2 + [18a_{13}(a_{31} - a_{13}) - 16a_{22}^2]\,Z + 9a_{13}^2 = 0. \tag{52}$$

The most degenerate case of Eq. (52) occurs when Z has a double root, which requires that the discriminant of (52) vanish, giving the condition:

$$a_{13} = \frac{4a_{22}^2}{9a_{31}} \tag{III}$$

in which case the double root Z becomes:

$$Z = \frac{4a_{22}^2}{9a_{31}^2 + 4a_{22}^2} \tag{53}$$

where Eq. (III) has been used to simplify the expression for Z. Note that $Z = \cos^2 \psi$ lies in between 0 and 1. Substituting Eq. (53) into Eq. (50) and using Eq. (III) to simplify the result gives:

$$\frac{\varDelta}{R^2} = \frac{24a_{22}^2 a_{40} - 9a_{22}a_{31}^2 - 54a_{04}a_{31}^2 + 4a_{22}^3}{2(9a_{31}^2 + 4a_{22}^2)}. \tag{IV}$$

Our conclusion is that in order that a system exhibit a maximally degenerate EO, its parameters must satisfy Eqs. (III) and (IV).

So in order to obtain those systems which simultaneously exhibit maximal degeneracies for both NNM's and EO's, we must find those points in the six dimensional parameter space $\varDelta/R^2$, $a_{40}, a_{31}, a_{22}, a_{13}, a_{04}$ which satisfy the bifurcation Eqs. (II), (III), (IV) and either (I a) or (I b). Each of these equations may be written as homogeneous polynomials in these parameters. This permits us to normalize the parameters, thereby reducing the number from 6 to 5 without loss of generality. We do this by dividing each parameter by a_{31}, i.e. by defining new parameters as:

$$\hat{a}_{ij} = \frac{a_{ij}}{a_{31}}, \qquad \hat{\varDelta} = \frac{\varDelta}{a_{31}}. \tag{54}$$

The resulting bifurcation equations will be the same as the ones we already have if we set

$$a_{31} = 1 \tag{55}$$

everywhere and replace the original parameters with the new ones. We do this, but for convenience we omit the ˆ notation. So now we need only deal with 5 parameters, but the bifurcation eqs. are no longer homogeneous in these parameters.

The number of parameters that we need to consider can be further reduced by using Eqs. (III) and (IV) to respectively eliminate a_{13} and $\varDelta/R^2$ from Eqs. (I a), (I b) and (II). We shall refer to the resulting eqs. as Eqs. (I a*), (I b*) and (II*) respectively. These are greatly simplified, and depend only on a_{22}, a_{40} and a_{04}. Inspection of the resulting expressions shows that all occurrences of a_{04} and a_{40} are in the form $a_{04} + a_{40}$, which for convenience we shall refer to as Q:

$$Q = a_{40} + a_{04}. \tag{56}$$

Moreover, Eq. (I b*) is able to be factored into two simpler equations, which we shall refer to as Eqs. (I b1*) and (I b2*). Here is the form of the resulting bifurcation equations:

$$256a_{22}^8 + 1024a_{22}^6 - 1536a_{22}^5 Q + 2304a_{22}^4 Q^2$$

$$+ 2016a_{22}^4 - 6912a_{22}^3 Q + 3888a_{22}^2 + 729 = 0, \tag{Ia*}$$

$$16a_{22}^4 - 144a_{22}^2 + 216a_{22}Q - 567 = 0, \tag{Ib1*}$$

$$16a_{22}^4 - 216a_{22}Q + 81 = 0, \tag{Ib2*}$$

$$[16a_{22}^4 - 216a_{22}Q + 81]\,[55296a_{22}^8 Q + 25 \text{ similar terms}] = 0. \tag{II*}$$

Note that Eq. (II*) factors into Eq. (I b2*) as well as a long polynomial. This means that any points (a_{22}, Q) which satisfy Eq. (I b2*) also satisfy Eq. (II*), and hence are points of maximum degeneracy. Such points lie on the following curve in the $Q - a_{22}$ plane, see Fig. 7:

$$Q = \frac{2a_{22}^3}{27} + \frac{3}{8a_{22}}. \tag{57}$$

Have we found all such points (a_{22}, Q) which exhibit maximum degeneracy? In order to find out, we must see if Eq. (II*) can be solved simultaneously with either of Eqs. (I a*) or (I b1*). We first eliminate Q between Eqs. (II*) and (I a*). a step which can be conveniently accomplished in MACSYMA using the ELIMINATE command, which results in a polynomial on a_{22} which turns out to have no real roots. In the case of Eq. (I b1*), however, additional degenerate points do exist. This may be seen by eliminating Q between Eqs. (II*) and (I b1*), which gives the following additional points of maximum degeneracy, see Fig. 7:

$$(a_{22}, Q) = \left\{ \left(\frac{3}{2}, \frac{5}{2} \right), \left(\frac{-3}{2}, \frac{-5}{2} \right) \right\}. \tag{58}$$

The parameter values of a_{13} and Δ/R^2 which correspond to the degenerate points of Eqs. (57), (58) are obtained by substituting into Eqs. (III), (IV).

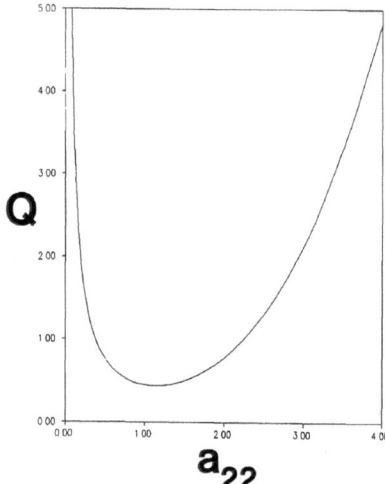

Fig. 7. Points of maximum degeneracy in $Q - a_{22}$ parameter space, see Eqs. (57) and (58). $Q = a_{40} + a_{04}$, Eq. (56)

6 Unfolding the singularity

We have shown that (to within the order of approximation of the perturbation method) there exist systems which exhibit maximally degenerate singularities. We now investigate systems which are close to these degenerate systems. The procedure involves prescribing a small but arbitrary change in parameters away from a degenerate point, in order to obtain all possible local bifurcations of PM's. The result is called a universal unfolding of the degenerate point [3].

Let a typical point on the curve (57) be parameterized by a_{22}^0. Then the corresponding value of $Q = a_{40}^0 + a_{04}^0$ is given by Eq. (57), while the values of a_{13}^0 and Δ^0/R^2 are given by Eqs. (III) and (IV). The value of a_{31}^0 remains normalized at unity. Corresponding to these parameters, there is a degenerate singularity in the $\varphi - \psi$ plane located at

$$\varphi^0 = \pi \quad \text{and} \quad \psi^0 = \arctan\left[\frac{3}{2a_{22}^0}\right] \tag{59}$$

(respectively from Eqs. (43) and (53)). (We omit the choice $\varphi^0 = 0$, $\psi^0 = -\arctan(3/[2a_{22}^0])$ since it gives an equivalent unfolding to that of Eqs. (59).) Thus we set:

$$a_{22} = a_{22}^0 + b_{22}, \quad a_{04} = a_{04}^0 + b_{04}, \quad a_{40} = a_{40}^0 + b_{40},$$

$$a_{13} = a_{13}^0 + b_{13}, \quad \frac{\Delta}{R^2} = \frac{\Delta^0}{R^2} + b_\Delta, \quad \varphi = \varphi^0 + u, \quad \psi = \psi^0 + v \tag{60}$$

where the b_{ij}'s are unfolding parameters (assumed small) and where u and v represent the associated displacement of the singular point from (φ^0, ψ^0). Our goal is to obtain expressions for u and v in terms of the b_{ij}'s.

We begin by substituting Eqs. (59), (60) into Eqs. (29), (30) for $f(\varphi, \psi)$ and $g(\varphi, \psi)$. This gives two Eqs. on u and v with the solution $u = v = 0$ when all the $b_{ij} = 0$. We separately investigate NNM's and EO's. The eq. that comes from $g(\varphi, \psi) = 0$ is satisfied by $u = 0$, corresponding to NNM's (which satisfy $\varphi \equiv \varphi^0$). Substituting $u = 0$ into $f(\varphi, \psi) = 0$, Eq. (29), and Taylor-expanding the result for small v gives:

$$K_0 + K_1 v + K_2 v^2 + K_3 v^3 + \cdots = 0 \tag{61}$$

where the K_i are known functions of the b_{ij}'s as well as the a_{ij}^0's. Since $f(\varphi, \psi)$ is linear in the a_{ij}'s, the K_i's are affine in the b_{ij}'s.

In order to similarly treat the EO's, we solve $g(\varphi, \psi) = 0$ for $\cos \varphi$ as in Eq. (44), substitute the result into $f(\varphi, \psi) = 0$, and obtain Eq. (50) for $\cos^2 \psi$. Substituting Eqs. (59) and (60) into Eq. (50) and Taylor-expanding for small v gives:

$$L_0 + L_1 v + \cdots = 0 \tag{62}$$

where the L_i are known functions of the b_{ij}'s as well as the a_{ij}^0's.

Conditions (58) on NNM's and (59) on EO's can be simplified without loss of generality by choosing b_Δ to make K_0 vanish and b_{13} to make L_0 vanish. Generality is assured because the results of the analysis will depend upon only two unfolding parameters (a "codimension 2" bifurcation). We obtain:

$$b_4 = \frac{3(32b_{40}a_{22}^{04} - 12b_{22}a_{22}^{02} - 72b_{04}a_{22}^{02} + 27b_{22})}{8a_{22}^{02}(4a_{22}^{02} + 9)}, \tag{63}$$

$$b_{13} = \frac{b_{22}(4a_{22}^{02} - 9)}{9a_{22}^0}. \tag{64}$$

With $K_0 = 0$, Eq. (58) has the root $v = 0$, as well as two additional possible roots which, if they exist, may be written in the approximate form

$$\text{NNM's:} \quad u = 0, \quad v^2 \sim \frac{27}{a_{22}^0[4a_{22}^{02} + 9]^4}$$

$$\times [(64a_{22}^{06} + 144a_{22}^{04} + 540a_{22}^{02} + 243)\, b_{22} - 576a_{22}^{04}(b_{04} + b_{40})]. \tag{65}$$

Eq. (62) with $L_0 = 0$ gives $v = 0$ for EO's. Substitution of this result and Eqs. (59), (60) into Eq. (51) and Taylor-expanding for small u gives an approximate expression for u in terms of the b_{ij}'s for EO's:

$$\text{EO's:} \quad v = 0, \quad u^2 \sim \frac{4a_{22}^{02} + 9}{4a_{22}^{03}}\, b_{22}. \tag{66}$$

Inspection of Eqs. (65), (66) shows that the unfolding parameters are naturally chosen as b_{22} and $(b_{04} + b_{40})$. Eq. (65) shows that in addition to the NNM at $u = v = 0$ (the degenerate singularity), two additional NNM's occur if

$$b_{22} > \frac{576a_{22}^{04}}{64a_{22}^{06} + 144a_{22}^{04} + 540a_{22}^{02} + 243}\, (b_{04} + b_{40}). \tag{67}$$

Similarly, Eq. (66) shows that an EO (and it's mirror image in the line $\varphi = \pi$) occur if

$$b_{22} > 0. \tag{68}$$

The straight lines (67) and (68) are bifurcation curves in the unfolding, see Fig. 8. Another bifurcation curve is obtained by considering the stability of the EO. This may be ascertained as follows: Let the first integral (31) be called $h(\varphi, \psi) = $ constant. Then at the EO singularity (66), h_φ and h_ψ vanish. The transition from stable to unstable occurs when the determinant of the

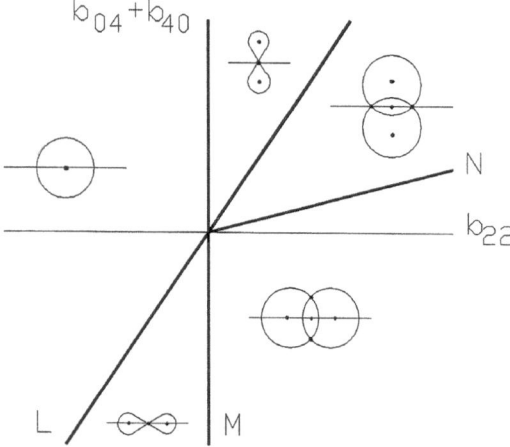

Fig. 8. Universal unfolding of a point of maximum degeneracy. The lines L, M and N respectively correspond to the conditions (67), (68) and (70). A phase portrait in the $\varphi - \psi$ plane is displayed in each region. The horizontal line in each phase portrait corresponds to $\varphi = \pi$. The phase portrait on line N consists of a circle of nonisolated singular points

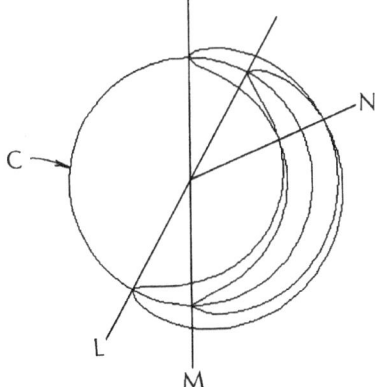

Fig. 9. Bifurcation diagram showing the location of singular points (PM's) which lie in the neighborhood of a point of maximum degeneracy. The circle C represents the maximally degenerate singularity, and the other curves represent singularities which have bifurcated from it as one moves in a circle around the origin in Fig. 8. The radial distance from the circle C represents the distance in the $\varphi - \psi$ phase plane from the maximally degenerate singularity to the bifurcated singularities. As in Fig. 8, the lines L, M and N respectively correspond to the conditions (67), (68) and (70)

Hessian vanishes:

$$\det \begin{bmatrix} h_{\varphi\varphi} & h_{\varphi\psi} \\ h_{\psi\varphi} & h_{\psi\psi} \end{bmatrix} = 0. \tag{69}$$

Substitution of Eqs. (59), (60) and (66) into (69) gives

$$b_{22} = \left[\frac{216 a_{22}^{02}}{16 a_{22}^{04} + 81} \right] (b_{04} + b_{40}). \tag{70}$$

Figure 8 shows the unfolding of a maximally degenerate singularity. The phase portraits displayed in Fig. 8 show the region of the $\varphi - \psi$ plane surrounding the degenerate singularity at $\varphi^0 = \pi$, and include a portion of the plane for which $\varphi > \pi$, which as we have noted previously, is a mirror image of $\varphi < \pi$. The phase portrait on the curve (70) involves a circle of nonisolated singular points.

If one were to traverse a circle around the origin in Fig. 8, the bifurcation diagram of Fig. 9 would be encountered. This may be described as two double pitchforks joined together by two transcritical bifurcations [3].

7 Conclusions

Previous studies of NNM's have shown how the number of NNM's may change through bifurcation [4, 10]. The present paper goes beyond previous work in several ways. To begin with, through the power of computer algebra, we have been able to study the general class of systems (1)−(7), rather than a specific system (cf. [4, 10]). For this class of systems, we have been able to obtain an approximate first integral which has permitted us to determine the number and stability of PM's. Moreover, for systems in the general class (1)−(7), we have been able to obtain the most degenerate possible bifurcation of PM's, and have obtained a universal unfolding which describes all possible local behavior of PM's. In particular, we have shown how NNM's are related to a family of PM's which we have called EO's. While NNM's project onto the $x - y$ configuration plane as line segments, EO's project as ellipses.

These results are seen as a first step in understanding the relationship between NNM's and more general classes of periodic orbits. It is expected that additional bifurcations exist in the class of systems considered here, and that approximate descriptions of such bifurcations may be obtained by extending the perturbation analysis to include higher order terms.

Acknowledgement

The authors wish to thank Professors John Guckenheimer, Timothy Healey and Philip Holmes of Cornell University for their helpful suggestions. The second author expresses his gratitude to the Korea Science and Engineering Foundation for partial support of this work.

References

[1] Bauer, L., Reiss, E. L., Keller, H. B.: Axisymmetric buckling of hollow spheres and hemispheres. Comm. Pure Applied Math. **23**, 529 – 568 (1970).

[2] Deprit, A., Henrard, J.: A manifold of periodic orbits. Advances in Astronomy and Astrophysics **6**, 1 – 124 (1968).

[3] Guckenheimer, J., Holmes, P. J.: Nonlinear oscillations, dynamical systems and bifurcations of vector fields. Berlin, Heidelberg, New York, Tokyo: Springer 1983.

[4] Johnson, T. L., Rand, R. H.: On the existence and bifurcation of minimal normal modes. Int. J. Nonlinear Mechanics **14**, 1 – 12 (1979).

[5] Kauderer, H.: Nichtlineare Mechanik. Berlin, Göttingen, Heidelberg, New York: Springer 1958.

[6] Kevorkian, J., Cole, J. D.: Perturbation methods in applied mathematics. Berlin, Heidelberg, New York: Springer 1981.

[7] Kinney, W., Rosenberg, R. M.: On steady state vibrations of nonlinear systems with many degrees of freedom. J. Appl. Mech. **33**, 406 – 412 (1966).

[8] Manevich, L., Mikhlin, I.: On periodic solutions close to rectilinear normal modes. PMM **36**, 1051 – 1058 (1972).

[9] Mikhlin, I.: Resonance modes of near-conservative systems. PMM **38**, 459 – 464 (1974).

[10] Month, L. A., Rand, R. H.: An application of the Poincare map to the stability of nonlinear normal modes. J. Appl. Mech. **47**, 645 – 651 (1980).

[11] Nayfeh, A. H.: Perturbation methods. New York: Wiley 1973.

[12] Pak, C. H.: On the stability behavior of bifurcated normal modes in coupled nonlinear systems. J. Appl. Mech. **56**, 155 – 161 (1989).

[13] Pak, C. H., Rosenberg, R. M.: On the existence of normal mode vibrations in nonlinear systems. Q. Applied Math. **26**, 403 – 416 (1968).

[14] Rand, R. H.: A higher order approximation for nonlinear normal modes in two degree of freedom systems. Int. J. Nonlinear Mechanics **6**, 545 – 547 (1971).

[15] Rand, R. H., Armbruster, D.: Perturbation methods, bifurcation theory and computer algebra. Berlin, Heidelberg, New York, Tokyo: Springer 1987.

[16] Rosenberg, R. M.: On nonlinear vibrations of systems with many degrees of freedom. Advances in Applied Mechanics **9**, 155 – 242 (1966).

[17] Shaw, S. W., Pierre, C.: Normal modes and superposition in nonlinear oscillatory systems. Preprint (1991).

[18] Szebehely, V.: Theory of orbits. New York: Academic Press 1967.

[19] Vakakis, A. F.: Analysis and identification of linear and nonlinear modes in vibrating systems. Ph. D. Thesis, California Institute of Technology, Pasadena, California 1990.

[20] Vakakis, A. F., Rand, R. H.: Normal modes and global dynamics of a two degree of freedom system. Submitted for publication to Int. J. Nonlinear Mechanics (1991).

Authors' address: Prof. R. H. Rand, Department of Theoretical and Applied Mechanics, Cornell University, Ithaca, NY 14853-1503, U.S.A. Prof. C. H. Pak, Department of Mechanical Engineering, Inha University, Inchon, 160, Korea. Prof. A. F. Vakakis, Department of Mechanical Engineering, University of Illinois, Urbana, Illinois, 61801, U.S.A.

Acta Mechanica (1992) [Suppl] 3: 147–160

Dynamic buckling delamination of a bonded thin film under residual compression

Y. J. **Lee** and **L. B. Freund**, Rhode Island, U.S.A.

Summary. A thin solid film or coating bonded to a solid substrate may be in a state of residual stress due to mismatch in thermal expansion coefficient between the film and the substrate or other effects. If the residual stress is compressive, the tendency for film buckling is suppressed by the relatively high stiffness of the substrate. However, at a flaw in the interfacial bonding between the film and substrate, buckling of the film can and does occur. The focus here is on the process of delamination buckling under these circumstances, including dynamic effects. For an interfacial defect of a certain size, the compressive force in the film may exceed the buckling load, in which case the buckling process is inherently dynamic. Both cases of plane strain and axially symmetric deformation are considered, and propagation of the buckle is permitted provided that an energy balance separation condition is satisfied. Post-bifurcation response of the film is described by means of the von Karman plate theory. Hamilton's principle is applied to obtain an approximate representation of the deformation in terms of two generalized coordinates, namely, the midpoint deflection of the buckled region and the size of the buckled region. Dynamic effects included are the transient deformation from the bifurcation state to the post-buckling configuration and the possibility of buckle nucleation due to waves impinging on the film from within the substrate. Histories of transient buckle deflection and buckle width are determined for representative material parameters.

1 Introduction

The configuration of a thin film bonded to a substrate is very common in microelectronic devices, composite materials and protective coating technologies. If the bonded film carries a residual compressive stress then it has the potential for delamination by buckling, and mechanics modeling provides a means to quantitatively characterize the process. This is essential to interpret experiments on buckling delamination from which interface toughness is to be inferred, to predict failure in systems with known material and strength properties, or to design a process for controlled delamination.

Significant progress has been reported in modeling delamination buckling under quasistatic deformation [1, 2, 3, 4]. In the simplest characterization of interface separation resistance in terms of a Griffith-like fracture energy per unit area, it is observed that spontaneous delamination which expands indefinitely is precipitated by a buckle if the interface bonding energy is sufficiently low. Once bifurcation from the initially flat configuration occurs, equilibrium can no longer be satisfied in such cases. The inclusion of inertial effects makes it possible to actually follow the expansion of the dilamination front under these conditions. In more recent work, Argon et al. [5] and Hutchinson et al. [6] have interpreted buckle shapes more complex than those considered here. These shapes are essentially circular but with waviness or wrinkling in the circumferential direction.

Recently, Gupta and Argon [7] conducted experiments in which they measured the strength of thin film interfaces by laser spalling. They induced a plane compressive stress pulse of short duration, parallel to the free surface of the film, which propagated through the substrate toward the thin film. This pulse was then reflected from the free surface of the film as a tensile stress pulse, providing an additional driving force for delamination. Even though they measured the strength of an interface with no perceptible flaws, the technique may be also used to measure the flaw toughness of the interface. Another dynamic test of interface strength in fiber reinforced composite materials was reported by Chai [8]. A preloaded composite panel was impacted by a projectile launched from an air gun. Macroscopic measurements and microscopic damage observations were reported.

In this paper, the dynamic propagation and arrest of a delamination buckle are studied by means of a plate model of the thin film, following the lead of earlier work. A one term Rayleigh-Ritz approximation of the deformation is adopted, and Hamilton's principle is used to determine the nonlinear differential equations governing the amplitude of the deformation $q(t)$ and the time-dependent size of the delamination $a(t)$. These differential equations are solved numerically, and implications for buckle propagation are considered.

2 Features of the model

The process of dynamic delamination buckling is precipitated by a flaw in the interface between the compressed thin film and the relatively stiff substrate. Here, the shape of the initial flaw is assumed to be either a long uniform strip, so that buckling occurs under plane strain conditions, or a circle, so that buckling occurs under conditions of axial symmetry. Under suitable conditions, the compressed film can lower its potential energy by transverse deflection, and this energy becomes available to accelerate the film as it deflects. The transverse deformation induces a stress concentration at the boundary of the region of delamination which may be sufficiently severe to cause this region to expand. As the delamination expands, additional elastic energy which was stored in the film as residual strain becomes available to accelerate the film or to drive the delamination process. It is the purpose here to devise a mathematical model of this process.

2.1 Plane strain deformation

Consider the configuration shown in Fig. 1. The film of thickness h is initially planar, and it carries a uniaxial compressive strain of magnitude ε_0 in the x-direction prior to buckling. Initially, the film is attached to the rigid substrate over $a_0 < |x| < \infty$ and it is free over $-a_0 < x < a_0$, the region of the defect. If the film deflects from this configuration with transverse

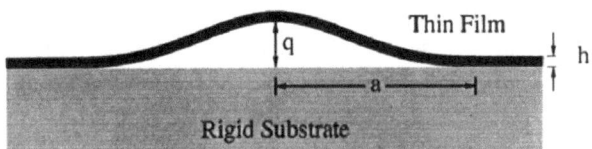

Fig. 1. Cross section of the buckled film attached to a rigid substrate for either plane strain or axially symmetric deformation, showing the midpoint deflection $q(t)$ and the buckle size $a(t)$

displacement $w(x, t)$ and axial displacement $u(x, t)$ then, according to the von Karman nonlinear plate theory [9], the extensional and bending strain energies of the film in $0 < x < l$ are

$$U_{\text{ext}} = \frac{Eh}{2(1 - v^2)} \left[\int_0^a \left(u_x + \frac{w_x^2}{2} \right)^2 dx + \varepsilon_0^2(l - a) \right] \tag{2.1}$$

and

$$U_{\text{bend}} = \frac{D}{2} \int_0^a w_{xx}^2 \, dx \tag{2.2}$$

where E is Young's modulus, v is Poisson's ratio, $D = Eh^3/12(1 - v^2)$ is the bending stiffness of the plate, $a(t)$ is the current half-width of the buckle, l is an arbitrary fixed length greater than a for all time of interest, and a subscript x denotes a derivative with respect to x. Because only changes in energy are of significance, the parameter l plays no role in the motion. The last term in (2.1) is retained, however, because it accounts for the energy being added to the buckled region from the initial residual strain as the buckle expands. The total potential energy, which is determined by the current deformation, is then

$$V_{\text{tot}} = U_{\text{ext}} + U_{\text{bend}}. \tag{2.3}$$

The total kinetic energy of the film is

$$T_{\text{tot}} = \frac{h}{2} \int_0^a \varrho(u_t^2 + w_t^2) \, dx \tag{2.4}$$

where ϱ is the material mass density and the subscript t denotes a time derivative. As an assumption with established validity, the contribution due to axial motion in the second term of equation (2.4) will be neglected in the analysis to follow [10].

A kinetically admissible deformed shape for the buckle which is expected to be reasonably accurate even under dynamic conditions is the shape corresponding to the fundamental equilibrium, buckling mode,

$$w(x, t) = \frac{1}{2} q(t) \left[1 + \cos\left(\frac{\pi x}{a(t)} \right) \right] \tag{2.5}$$

and

$$u(x, t) = -\varepsilon_0 x + \frac{\pi}{32} \frac{q(t)^2}{a(t)} \sin\left(\frac{2\pi x}{a(t)} \right) \tag{2.6}$$

where $a(t)$ and $q(t)$ are interpreted as the generalized coordinates characterizing the configuration of the system. For any given $a(t)$ and h, the critical strain at which buckling will occur according to the equilibrium theory is given by

$$\varepsilon_b = \frac{\pi^2}{12} \frac{h^2}{a^2}. \tag{2.7}$$

A parameter useful for comparison of the severity of the loading for a defect of a given size a is

$$m = \frac{\varepsilon_0}{\varepsilon_b} = \frac{12a^2\varepsilon_0}{\pi^2 h^2}. \tag{2.8}$$

The particular value of m when $a = a_0$ is denoted by m_0. Clearly, buckling will not occur at all unless $m_0 \geqq 1$.

With the motion of the system described in this way, Hamilton's principle of least action requires that

$$\frac{d}{dt} \frac{\partial L}{\partial q_t} - \frac{\partial L}{\partial q} = 0,$$

$$\frac{d}{dt} \frac{\partial L}{\partial a_t} - \frac{\partial L}{\partial a} = -G \tag{2.9}$$

where L is the Lagrangian function $L = T_{\text{tot}} - V_{\text{tot}}$ [11, 12]. The dynamic energy release rate G is the "dissipative" generalized force which is work-conjugate to $a(t)$. The equations of motion written in the form (2.9) provide two constraints on the three functions $q(t)$, $a(t)$ and $G(t)$, so the system of equations implied by (2.9) is underdetermined. However, if an interface separation condition is added to these equations then the system of equations becomes consistent, in the sense that the number of equations becomes equal to the number of unknowns. The separation condition used here requires that either $a_t = 0$ with $G < G_c$ or $G = G_c$ with $a_t > 0$. Although G_c is assumed to be constant for this calculation, it is likely that the value of G_c depends on the local phase of the state of stress in the material at the edge of the delamination front [13].

As a brief aside, it is a relatively simple matter to reduce the second equation in (2.9) to the more common form

$$a_t G = -\frac{d}{dt} (T_{\text{tot}} + V_{\text{tot}}) \tag{2.10}$$

for dynamic energy release rate [12]. Essentially, this reduction requires that V_{tot} should be independent of a_t, q_t and that T_{tot} should be a homogeneous function of degree two in these same variables [14].

At this point, it is convenient to introduce the nondimensional variables defined by

$$\bar{q} = \frac{q}{h}, \qquad \bar{a} = \frac{a}{h}, \qquad \bar{t} = \frac{c_p t}{h}, \tag{2.11}$$

$$\bar{G} = \frac{G}{Eh/(1 - v^2)}, \qquad \bar{G}_c = \frac{G_c}{Eh/(1 - v^2)}, \tag{2.12}$$

$$\bar{V}_{\text{tot}} = \frac{V_{\text{tot}}}{Eh^2/(1 - v^2)}, \qquad \bar{T}_{\text{tot}} = \frac{T_{\text{tot}}}{Eh^2/(1 - v^2)} \tag{2.13}$$

where $c_p = \sqrt{E/\varrho(1 - v^2)}$ is the extensional wave speed of the plate, G is the energy release rate at each end of the zone of delamination and G_c is the critical energy release rate for separation of the material interface. From this point onward, a superposed dot on a symbol will be used to denote its derivative with respect to normalized time $\bar{t}$. With these definitions, the total potential energy of the system is

$$\bar{V}_{\text{tot}} = \frac{\pi^4 \bar{q}^4}{512 \bar{a}^3} - \frac{\pi^2}{16} \frac{\bar{q}^2}{\bar{a}} \left(\varepsilon_0 - \frac{\pi^2}{12 \bar{a}^2} \right) \tag{2.14}$$

and the kinetic energy is

$$\bar{T}_{\text{tot}} = \frac{3}{16}\,\bar{a}\dot{\bar{q}}^2 + \frac{3}{16}\,\bar{q}\dot{\bar{a}}\dot{\bar{q}} + \left(\frac{\pi^2}{48} - \frac{1}{32}\right)\frac{\bar{q}^2\dot{\bar{a}}^2}{\bar{a}}. \tag{2.15}$$

If $m_0 \geq 1$, the film begins to deflect with zero initial speed. As the potential energy decreases, conservation of energy $\bar{V}_{\text{tot}} + \bar{T}_{\text{tot}}$ requires that the kinetic energy increases according to

$$\dot{\bar{q}} = \frac{2\varepsilon_0\bar{q}}{m_0}\sqrt{m_0 - 1 - \frac{3}{8}\,\bar{q}^2}. \tag{2.16}$$

If there is no further delamination, the deflection of the film passes through an equilibrium position. It then decelerates and comes to rest at some deflection $\bar{q}_{\max}$ which, according to (2.16), is

$$\bar{q}_{\max} = 2\sqrt{2(m_0 - 1)/3} \tag{2.17}$$

and a vibration about some deflected equilibrium ensues.

The energy release rate $\bar{G}$ can be calculated directly from (2.12). Prior to further delamination, $a = a_0$ and $a_t = 0$ of course. If $\dot{\bar{q}}$ is eliminated from the expression for $\bar{G}$ by means of (2.16), then an expression for $\bar{G}$ in terms of $\bar{q}$ is obtained. If the value of $\bar{q}$ when $\bar{G} = \bar{G}_c$ is denoted by $\bar{q}_c$ then it is found that

$$\bar{q}_c = \sqrt{\frac{8}{15}}\sqrt{(m_0 - 2) + \sqrt{\frac{5}{2}\frac{m_0^2}{\varepsilon_0^2}\,\bar{G}_c + (m_0 - 2)^2}}. \tag{2.18}$$

Thus, the zone of delamination will begin to expand during the dynamic buckling process provided that the value of $\bar{q}_{\max}$ determined from (2.17) is greater than the value of $\bar{q}_c$ determined from (2.18). If further delamination does not occur, the present problem is reduced to a relatively simple nonlinear, one degree-of-freedom vibration problem. Thus, attention is focussed on the case of delamination growth in the next section.

As a base of reference, the equilibrium energy release rate can be derived by the above procedure simply by neglecting the kinetic energy contribution. The result is

$$\bar{G}_{eq} = \bar{\Gamma}_0\left(1 - \frac{1}{m}\right)\left(1 + \frac{3}{m}\right) \tag{2.19}$$

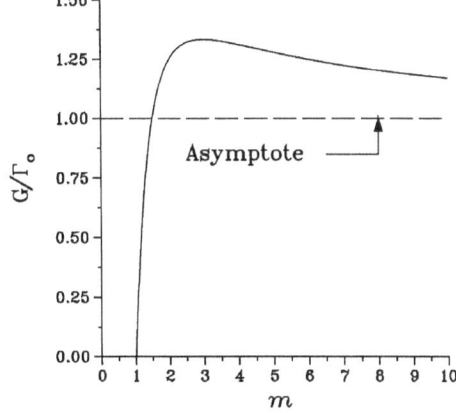

Fig. 2. Normalized energy release rate versus load factor m under equilibrium conditions for the case of plane strain deformation

where $\bar{\Gamma}_0$ denotes the normalized saturation energy release rate which is the limiting result as m increases indefinitely. Its expression in terms of system parameters is simply

$$\bar{\Gamma}_0 = \frac{\varepsilon_0{}^2}{2}. \tag{2.20}$$

This saturation energy release rate may be easily derived by direct application of the J-integral of elastic fracture mechanics.

The dependence of the equilibrium energy release rate on mismatch strain ε_0 and delamination length a is shown in Fig. 2. For constant values of the critical delamination energy G_c, three distinct ranges of response can be identified in Fig. 2:

(i) if $0 < G_c/\Gamma_0 \leq 1$ and $G_{eq}(m_0) > G_c$ then the flaw will expand spontaneously to result in complete failure of the interface.

(ii) if $1 < G_c/\Gamma_0 < \dfrac{4}{3}$ and $G_{eq}(m_0) > G_c$ then the flaw will grow stably to a size m for which $G_{eq}(m) > G_c$, whereupon expansion will be arrested.

(iii) if $G_{eq}(m_0) < G_c$ then the interface is sufficiently tough to prevent any expansion of the defect, even though buckling may occur.

2.2 Axially symmetric deformation

For the case of a circular defect in the interface bonding between a biaxially compressed film and the substrate, it is assumed that a buckle develops in an axisymmetric mode. The compressive residual strain is again denoted by ε_0 and the radius of the buckle is a. Thus, Fig. 1 again provides a schematic of the system as a section across a diameter.

The procedure for deriving equations of motion for this case follows that for the plane strain case. The extensional and bending energies of a circular elastic plate clamped along its boundary $r = a$, undergoing axially symmetric deformation represented by the transverse displacement $w(r, t)$ and the radial displacement $u(r, t)$, are

$$U_{\text{ext}} = \frac{\pi E h}{(1 - v^2)} \left[\int_0^a \left[\left(u_r + \frac{w_r{}^2}{2} \right)^2 + 2v \left(u_r + \frac{w_r{}^2}{2} \right) \frac{u}{r} + \left(\frac{u}{r} \right)^2 \right] r \, dr + (1 + v) \, \varepsilon_0{}^2 (l^2 - a^2) \right] \tag{2.21}$$

and

$$U_{\text{bend}} = \pi D \int_0^a \left[w_{rr}^2 + 2v w_{rr} \left(\frac{w_r}{r} \right) + \left(\frac{w_r}{r} \right)^2 \right] r \, dr \tag{2.22}$$

where the subscript r denotes partial differentiation with respect to the radial coordinate r. The parameter l is again an arbitrary length which is greater than a for all times of interest. With this interpretation, $V_{\text{tot}} = U_{\text{ext}} + U_{\text{bend}}$ is always the total potential energy within $0 < r < l$. If the in-plane contribution to kinetic energy is neglected in comparison to the transverse contribution, then

$$T_{\text{tot}} = \pi h \int_0^a \varrho w_t{}^2 r \, dr. \tag{2.23}$$

Hamilton's principle is applied to derive a pair of ordinary differential equations governing the deflection $q(t)$ of the center of the plate and the current size $a(t)$ of the delamination. Adapting a result from the work of Thompson and Hunt [15] to the present case, it is assumed that

$$w(r, t) = q \frac{J_0(\gamma) - J_0(\gamma r/a)}{J_0(\gamma) - 1} \tag{2.24}$$

and

$$u(r, t) = -\varepsilon_0 r + \frac{q^2}{a} \frac{\gamma}{4[J_0(\gamma) - 1]^2}$$
$$\times \left((1 + v) J_0(\gamma r/a) J_1(\gamma r/a) + v\gamma r/a[J_0(\gamma)^2 - J_0(\gamma r/a)^2 - J_1(\gamma r/a)^2]\right) \tag{2.25}$$

where J_n is the Bessel function of the first kind of integer order n and γ is the smallest positive root of $J_1(\gamma) = 0$, approximately $\gamma = 3.83$. The value of biaxial mismatch strain in the flat plate at which bifurcation can occur is

$$\varepsilon_b = \frac{\gamma^2}{12(1 + v)} \frac{h^2}{a^2}. \tag{2.26}$$

The differential equations governing the generalized coordinates $q(t)$ and $a(t)$ are given by Lagrange's equations

$$\frac{d}{dt} \frac{\partial L}{\partial q_t} - \frac{\partial L}{\partial q} = 0,$$
$$\frac{d}{dt} \frac{\partial L}{\partial a_t} - \frac{\partial L}{\partial a} = -2\pi a G \tag{2.27}$$

where $L = T_{tot} - V_{tot}$. As in (2.12), G denotes the energy released per unit length along the edge of the delamination.

It is not possible to extract the ordinary differential equations for $q(t)$ and $a(t)$ in the same simple way as for the plane strain case. Instead, the coefficients of the various terms in the energy expressions have been evaluated numerically with $v = 0.3$. Using normalizing factors similar to those introduced in (2.13), it is found that

$$\bar{V}_{tot} = 0.610 \frac{\bar{q}^4}{\bar{a}^2} - 2.47 \left(\varepsilon_0 - 0.940 \frac{1}{\bar{a}^2}\right) \bar{q}^2, \tag{2.28}$$

$$\bar{T}_{tot} = 0.259\bar{a}^2\dot{\bar{q}}^2 + 0.518\bar{q}\bar{a}\dot{\bar{q}}\dot{\bar{a}} + 0.634\bar{q}^2\dot{\bar{a}}^2. \tag{2.29}$$

For a given critical energy release rate G_c and initial flaw size a_0, the deflection at which $\bar{G} = \bar{G}_c$ is satisfied is

$$\bar{q}_c = 0.797 \sqrt{(m_0 - 2) + \sqrt{3.76 \frac{m_0^2}{\varepsilon_0^2} \bar{G}_c + (m_0 - 2)^2}}. \tag{2.30}$$

From the conservation of energy, the transverse velocity of the film during buckling is given in terms of m_0 and $\bar{q}$ by

$$\dot{\bar{q}} \approx 3.19\bar{q} \frac{\varepsilon_0}{m_0} \sqrt{(m_0 - 1) - 0.262\bar{q}^2}. \tag{2.31}$$

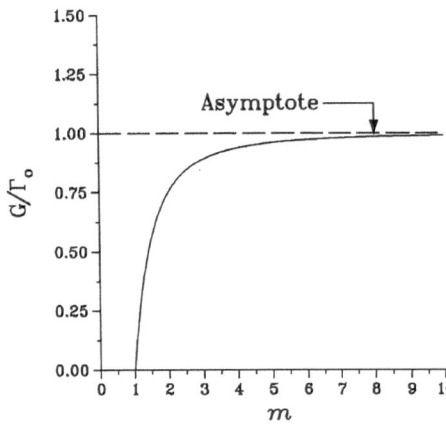

Fig. 3. Normalized energy release rate versus load factor m under equilibrium conditions for the case of axially symmetric deformation

From this result, it is evident that the maximum deflection is

$$\bar{q}_{max} \approx 1.95 \sqrt{m_0 - 1}. \tag{2.32}$$

The equilibrium energy release rate is

$$\bar{G}_{eq} = \bar{\Gamma}_0 \left(1 - \frac{1}{m^2} \right) \tag{2.33}$$

where the saturation energy release rate for the circular defect is [3]

$$\bar{\Gamma}_0 = 0.797 \varepsilon_0{}^2. \tag{2.34}$$

A graph of the equilibrium energy release rate is shown in Fig. 3. Its behavior is simpler than the corresponding plane strain result. If $0 < G_c/\Gamma_0 < 1$ and $G_{eq}(m_0) > G_c$ then delamination will occur spontaneously. Otherwise, delamination will not occur at all, according to the equilibrium theory.

3 Buckling and dynamic delamination

The equilibrium conditions under which buckling will result in further delamination of the film were established by Evans and Hutchinson [3], and they are evident from the discussion of Figs. 2 and 3. Essentially, it is required that $m_0 > 1$ and $\bar{q}_c \leqq \bar{q}_{eq}$, where $\bar{q}_{eq}$ is an equilibrium deflection. When dynamic effects are taken into consideration, there are additional circumstances under which further delamination can be induced in a buckling mode. One such circumstance arises when the initial size of the flaw a_0 is greater than the minimum flaw size which will result in buckling at a given mismatch strain ε_0, or $m_0 > 1$. In this case, the potential energy of the film decreases as it deflects toward an equilibrium position, as shown in Fig. 4, and the kinetic energy increases. The deflection passes through the equilibrium position (the minimum in Fig. 4) and it continues to deflect until the velocity is reduced to zero by the resistance to further deformation. (This is an example of the phenomenon of dynamic overshoot.) Thus, even though the equilibrium deflection may not be sufficiently large to induce delamination, the maximum deflection given in (2.17) or (2.32) may exceed the critical value (2.18) or (2.30), in which case delamination would be induced. In short, the

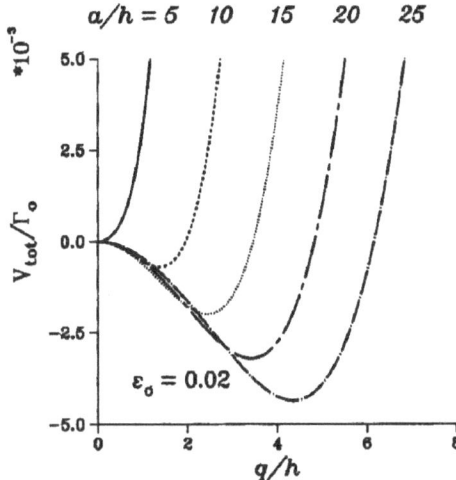

Fig. 4. Normalized potential energy versus transverse deflection for several values of the normalized buckle length a/h with $\varepsilon_0 = 0.02$

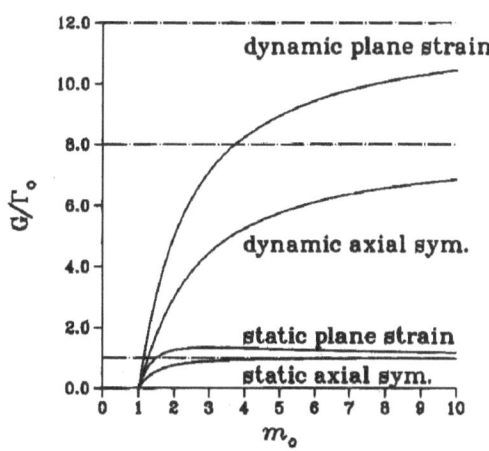

Fig. 5. Normalized dynamic energy release rate versus initial value of the load factor at the instant when the delamination growth condition is first satisfied, for both plane strain and axially symmetry

conditions for this to happen are

$$m_0 > 1 \quad \text{and} \quad \bar{q}_c \leqq \bar{q}_{\max}. \tag{3.1}$$

In terms of the initial value of $m_0 > 1$, the energy release rate calculated from (2.9) or (2.27) with $a = a_0$ has increased to the value

$$\bar{G} = 12\bar{\Gamma}_0 \left(1 - \frac{1}{m_0}\right)\left(1 - \frac{1}{3m_0}\right) \quad \text{(plane strain)} \tag{3.2}$$

or

$$\bar{G} = 8\bar{\Gamma}_0 \left(1 - \frac{1}{m_0}\right)\left(1 - \frac{1}{2m_0}\right) \quad \text{(axial symmetry)} \tag{3.3}$$

when $\bar{q}$ has increased to its maximum value $\bar{q}_{\max}$. These functions are shown graphically in Fig. 5. Suppose, for example, that a compressed film buckles under plane strain conditions with the length of the buckle being about $\sqrt{2}$ times the minimum length of the defect necessary for buckling to occur. Then $m = 2$ and the maximum energy release rate that can be achieved dynamically is about four times greater than the corresponding equilibrium value.

During growth of the delamination, $G = G_c$ and Lagrange's equations (2.9) provide two coupled ordinary differential equations for $\bar{q}(\bar{t})$ and $\bar{a}(\bar{t})$. Prior to the onset of growth of the defect, the length is a_0 and the motion is readily analyzed by means of conservation of energy. Thus, time $t = 0$ is chosen to be the instant at which the delamination condition $G = G_c$ is first satisfied. The initial conditions are then

$$\bar{q}(0) = \bar{q}_c, \qquad \dot{\bar{q}}(0) = \dot{\bar{q}}_c,$$

$$\bar{a}(0) = \bar{a}_0, \qquad \dot{\bar{a}}(0) = 0. \tag{3.4}$$

The system of ordinary differential equations is solved numerically by means of a conventional fourth-order Runge-Kutta method.

A representative result for plane strain buckling is shown in Fig. 6 for the case when $\varepsilon_0 = 0.02$ and $\bar{a}_0 = 10$, which implies that $m_0 = 2.43$. Results are shown for three values of the critical fracture energy, $\bar{G}_c/\bar{\Gamma}_0 = 0.5, 2, 4$. A unit on the nondimensional time axis is essentially the time required for a longitudinal wave to propagate a distance equal to the film thickness h. The period of free transverse vibration of the buckled portion of the film would be at least two or three orders of magnitude larger than this time for and $a \gg h$.

In all three cases in Fig. 6, the defect began to expand at $t = 0$. However, the three responses are quite different in detail. For the toughest interface with $\bar{G}_c/\bar{\Gamma}_0 = 4$, growth begins shortly before the deflection $\bar{q}(\bar{t})$ reaches a local maximum value, and the growth is arrested shortly thereafter with $\bar{a} \sim 11$. The amplitude of the subsequent vibration is insufficient to re-initiate the growth process. The interface is too tough for spontaneous delamination to occur. Indeed, it is too tough to permit any delamination under equilibrium conditions. For the case of intermediate toughness $\bar{G}_c/\bar{\Gamma}_0 = 2$, the delamination front accelerates to a fairly high speed, it then decelerates nearly to rest, and then undergoes a second acceleration/deceleration cycle before coming to rest with $\bar{a} \sim 15$. For the case of the lowest of the three toughness values illustrated $\bar{G}_c/\bar{\Gamma}_0 = 0.5$, spontaneous delamination is possible and the graph in Fig. 6 illustrates the dynamics of the nonequilibrium process. The moving delamination front rapidly accelerates up to a speed of roughly $0.075c_p$ and it propagates at this speed thereafter. It should be noted that this model is based on the assumption of instantaneous equilibration of extensional stresses in the film, and the validity of this approximation diminishes as the length of the delamination zone becomes very large.

Similar results are shown in Fig. 7 for the corresponding case of axially symmetric buckling. Again, $\varepsilon_0 = 0.02$ and $\bar{a}_0 = 10$, which implies that $m_0 = 2.13$. Results are shown for three values of the interface toughness $\bar{G}_c/\bar{\Gamma}_0 = 0.5, 1.5, 3.0$. Growth of the delamination front occurs for all three values of toughness, but the amount of growth is slight for the two larger values. For the lowest value of toughness, growth is sustained indefinitely at a speed similar to that observed in the case of plane strain deformation.

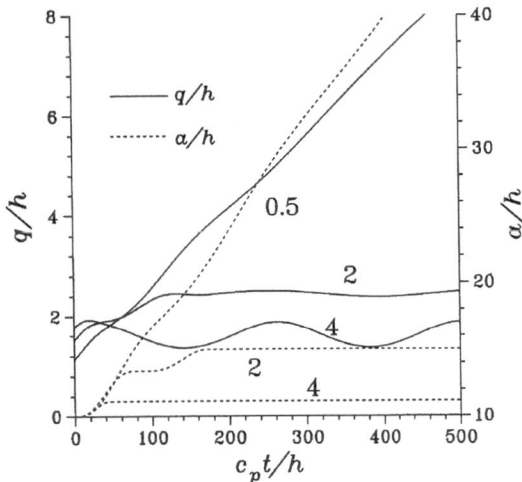

Fig. 6. History of the transverse deflection and buckle length during dynamic buckling delamination under plane strain conditions, with $\varepsilon_0 = 0.02$, $\bar{a}_0 = 10$, and $\dot{\bar{q}}_0 = 0$, for three values of interface toughness $\bar{G}_c/\bar{\Gamma}_0 = 0.5, 2$ and 4

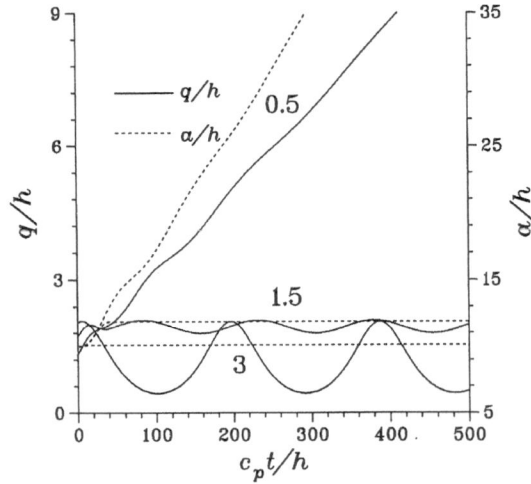

Fig. 7. History of the transverse deflection and buckle length during dynamic buckling delamination under conditions of axially symmetry, with $\varepsilon_0 = 0.02$, $\bar{a}_0 = 10$, and $\dot{\bar{q}}_0 = 0$, for three values of interface toughness $\bar{G}_c/\bar{\Gamma}_0 = 0.5, 1.5$ and 3

4 Impulsive loading with dynamic delamination

If the film is given an initial transverse velocity in the undeflected configuration, then it will begin to deflect even though the size of the defect may be less than the critical size for buckling to occur. Such an initial velocity can be imparted to the film by means of a wave incident from within the substrate. Suppose that a plane compressive stress pulse of length less than $2h$ impinges on the film from within the substrate. Because it is compressive, it will pass through the interface even though the interface may contain defects. The wave reflects from the free surface of the film as a tensile wave. In regions where the interface is not defective, the tensile wave can pass back into the substrate (assuming that the acoustic impedances of the two materials are not too different). In regions where the interface is defective, however, the tensile wave cannot pass back into the substrate. The momentum of the wave is trapped within the film, and the effect is to impart a transverse velocity to the film whose magnitude is determined by material properties and the amplitude of the incident wave. If σ^∞ is the amplitude of the incident stress wave, then the initial velocity of the film will be about $2\sigma^\infty/\sqrt{E\varrho}$. The nondimensional initial velocity is denoted by $\dot{q}_0$.

Again, if conservation of energy is applied to determine the maximum deflection due to the imposed initial velocity and the energy release rate is evaluated at the maximum deflection, it is found that

$$\frac{\bar{G}}{\bar{\Gamma}_0} = 6\left(1 - \frac{1}{3m_0}\right)\left[\left(1 - \frac{1}{m_0}\right) + \left|1 - \frac{1}{m_0}\right|\sqrt{1 + \frac{3}{8}\left(\frac{\dot{q}_0}{\varepsilon_0}\right)^2\left(\frac{m_0}{m_0 - 1}\right)^2}\right] + \frac{15}{8}\left(\frac{\dot{q}_0}{\varepsilon_0}\right)^2, \quad (4.1)$$

$$\frac{\bar{G}}{\bar{\Gamma}_0} = 4\left(1 - \frac{1}{2m_0}\right)\left[\left(1 - \frac{1}{m_0}\right) + \left|1 - \frac{1}{m_0}\right|\sqrt{1 + 0.103\left(\frac{\dot{q}_0}{\varepsilon_0}\right)^2\left(\frac{m_0}{m_0 - 1}\right)^2}\right] + 0.310\left(\frac{\dot{q}_0}{\varepsilon_0}\right)^2$$

$$(4.2)$$

for the cases of plane strain and axially symmetric deformation, respectively. These results are shown in Fig. 8 for several values of the nondimensional initial velocity $\dot{q}/c_p\varepsilon_0$. In these figures, the deflections are stable if $m_0 < 1$ and unstable if $m_0 > 1$.

Representative results for the case of impulsive loading and plane strain deformation are shown in Fig. 9 with $\varepsilon_0 = 0.02$, $\bar{a}_0 = 10$ and $\bar{G}_c/\bar{\Gamma}_0 = 2$. Expansion of the region of delamination is considered for three values of initial velocity $\dot{q}_0/\varepsilon_0 = 0, 0.5, 1$. The toughness of the interface is

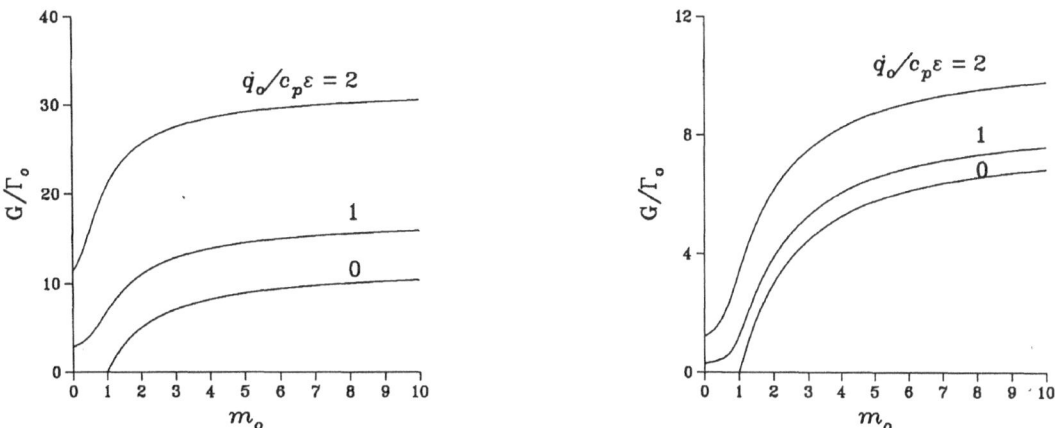

Fig. 8. Vales of the initial transverse velocity of the film induced by remote loading which are sufficient to result in dynamic film delamination versus the initial load factor m_0 for (a) the plane strain deformation and (b) axially symmetric deformation

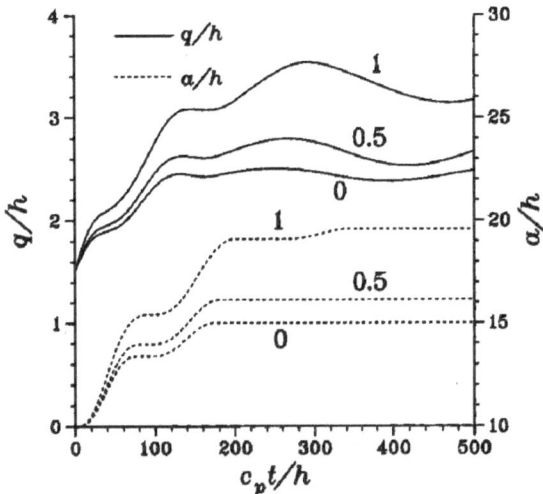

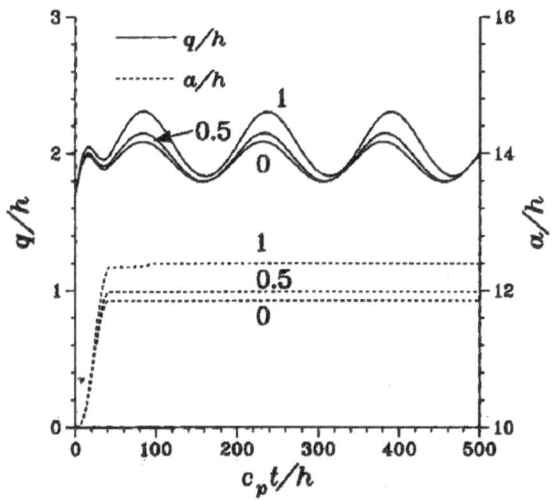

Fig. 9. History of transverse displacement and buckle size under plane strain conditions for the case of transverse velocity of the film induced by remote loading, with $\varepsilon_0 = 0.02$, $\bar{a}_0 = 10$, and $\bar{G}_c/\bar{\Gamma}_0 = 0.5$, for the three cases $\dot{\bar{q}}_0/\varepsilon_0 = 0$, 0.5 and 1

Fig. 10. History of transverse displacement and buckle size under axially symmetric conditions for the case of transverse velocity of the film induced by remote loading, with $\varepsilon_0 = 0.02$, $\bar{a}_0 = 10$, and $\bar{G}_c/\bar{\Gamma}_0 = 0.5$, for the three cases $\dot{\bar{q}}_0/\varepsilon_0 = 0$, 0.5 and 1

sufficiently great so that no expansion of the defect would occur under equilibrium conditions. Figure 9 shows that the buckle begins to expand at $t = 0$ in each case, and the delamination front then comes to rest as the film rebounds. Further expansion of the delamination zone may occur in the second and subsequent oscillations of the film, depending on the intensity of the initial impulse delivered to the film. Similar results are illustrated for the case of axially symmetric deformation in Fig. 10.

In Fig. 11, graphs of the total energy and the minimum total potential energy for the case of impulsive loading and plane strain deformation with $\varepsilon_0 = 0.02$, $\bar{a}_0 = 10$, $\bar{G}_c/\bar{\Gamma}_0 = 0.5$, and $\dot{\bar{q}}_0/\varepsilon_0 = 1$, are shown by a solid and a dashed line. The purpose of this figure is to predict an upper bound on the delaminated zone size prior to solving the equations of motion. This is accomplished by finding an intersection of the total energy and the minimum total potential energy curves, where the total kinetic energy vanishes. The computed total potential energy is also shown as a dotted line and the total kinetic energy is the difference between the total energy

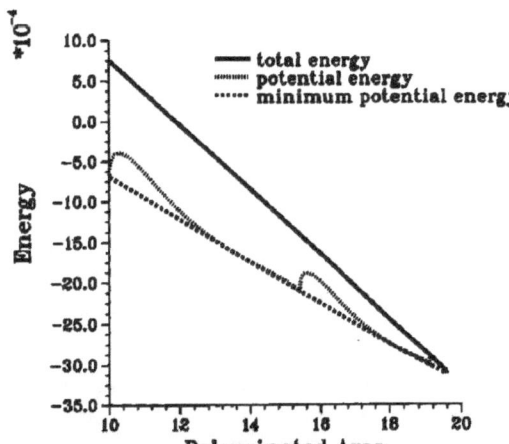

Fig. 11. The variation of total energy $\bar{V}_{tot} \cdot \bar{T}_{tot}$, potential energy $\bar{V}_{tot}$, and minimum potential energy with respect to deflection versus the size of delaminated zone a/h for plane strain, with $\varepsilon_0 = 0.02$, $\bar{a}_0 = 10$, $\bar{G}_c/\bar{\Gamma}_0 = 0.5$, and $\dot{\bar{q}}_0/\varepsilon_0 = 1$

and the total potential energy curve. It is interesting that the total potential energy curve is parallel to the total energy in a certain range, which implies a small variation of the total kinetic energy during delamination in this range.

5 Conclusion

Several observations follow from the discussion of the foregoing sections. If the conditions of a compressed thin film at the instant of bifurcation are supercritical, in the sense that the size of the interfacial defect is greater than the minimum size required for buckling to occur, then the buckling occurs dynamically. Under these conditions it is more likely that this dynamic buckling process will lead to further delamination of the interface than in the corresponding equilibrium case. This supercritical state might be reached, for example, if the interface defect is formed by corrosion or some other chemical process.

Under equilibrium conditions with a spatially uniform interface toughness independent of the phase angle of the local stress field, once delamination propagation begins it will continue indefinitely. However, in the dynamic case, it is possible to have the interface delamination zone extend some amount and then come to rest, provided that the toughness of the interface is in the appropriate range.

Even when the initial state of a compressed film with an interface flaw is subcritical, further delamination can be induced by means of a stress pulse incident on the film from within the substrate. For a short pulse, the film over a defect acts like a momentum trap, and the momentum trapped in the film in this way tends to lift the film off the substrate and to extend the zone of delamination. An incident stress wave can lead to spontaneous delamination of the entire film under certain conditions, even though the initial flaw size may have been subcritical.

Acknowledgement

The work described here was supported by the Office of Naval Research through contract N00014-90-J-4051 with Brown University. This research support is gratefully acknowledged.

References

[1] Chai, H., Babcock, C. D., Knauss, W. G.: One dimensional modeling of failure in laminated plates by delamination buckling. Int. J. Solids Struc. **17**, 1069–1083 (1981).

[2] Bottega, W. J., Maewal, A.: Delamination buckling and growth in laminates. J. Appl. Mech. **50**, 184–189 (1983).

[3] Evans, A. G., Hutchinson, J. W.: On the mechanics of delamination and spalling in compressed films. Int. J. Solids Struc. **20**, 455–466 (1984).

[4] Yin, W.-L., Fei, Z.: Delamination buckling and growth in a clamped circular plate. AIAA Journal **26**, 438–445 (1988).

[5] Argon, A. S., Gupta, V., Landis, H. S., Cornie, J. A.: Intrinsic toughness of interfaces between SiC coatings and substrates of Si or C fibre. J. Matls. Sci. **24**, 1207–1218 (1989).

[6] Hutchinson, J. W., Thouless, M., Liniger, E. G.: Growth and configurational stability of circular buckling-driven film delaminations. Harvard University Technical Report (1991).

[7] Gupta, V., Argon, A. S.: Measurement of strength of thin film interfaces by laser spallation experiment. Proceedings of the IUTAM Symposium on Inelastic Deformation of Composite Materials, to appear (1991).

[8] Chai, H.: The growth of impact damage in compressively loaded laminates. Ph. D. Thesis, California Institute of Technology (1982).

[9] Langhaar, H.: Energy methods in applied mechanics. New York: Wiley 1962.

[10] Mettler, E.: Dynamic buckling. In: Handbook of Engineering Mechanics (Flugge, W., ed.). New York: McGraw-Hill 1962.

[11] Burns, S. J., Webb, W. W.: Fracture surface energies and dislocation processes during dynamic cleavage of LiF. I. Theory. J. Appl. Phys. **41**, 2078 – 2085 (1970).

[12] Freund, L. B.: Dynamic fracture mechanics. Cambridge: Cambridge University Press 1990.

[13] Hutchinson, J. W., Suo, Z.: Mixed mode cracking in layered materials. In: Advances in Applied Mechanics **28** (Hutchinson, J. W., Wu, T. Y., eds.). New York: Academic Press, to appear (1992).

[14] Landau, L. D., Lifshitz, E. M.: Mechanics, 3rd ed. New York: Pergamon Press 1976.

[15] Thompson, J. M. T., Hunt, G. W.: A general theory of elastic stability. New York: Wiley 1973.

Authors' address: Prof. L. B. Freund, Division of Engineering, Brown University, Providence, RI 02 912, U.S.A.

Acta Mechanica (1992) [Suppl] 3: 161–171

Non-associated plastic deformation and genuine instability

D. C. Drucker and **M. Li**, Gainesville, Florida, U.S.A.

Summary. A non-associated flow rule often is employed to model the time-independent (elastic-plastic) behavior of materials sensitive to normal stress on planes of shear. To match the more moderate volume expansion observed, the plastic strain increment is taken as not normal to the current Mohr-Coulomb or similar yield surface. Unless the incremental elastic response were to override the plastic, however, there would be a genuine but limited instability of configuration for any such model subjected to a triaxial compression test. In fact, for the usual simple models, every state of stress in the plastic domain is unstable. The instability exhibited here is in the form of a rotating shear band. It is an initially accelerating instability, of limited total excursion, satisfying kinematics and kinetics. This instability of configuration would occur much earlier and differs fundamentally from the instability of the path of deformation customarily determined from the bifurcation condition for shear band formation under homogeneous stress and stress rate.

1 Introduction

Several types of behavior are found under the labels of bifurcation or instability in the elastic and the elastic-plastic domains. A bifurcation may be stable as in the departure from straightness in the plastic range of a perfect column under increasing axial load at or above the tangent modulus load but below the reduced modulus load. It was Shanley [1] who called our attention to this instability of path despite the stability of each straight configuration of the column. Bifurcation may be neutral, as for an elastic column at the Euler buckling load, or strongly unstable as in the buckling of a perfect cylindrial shell under axial stress.

Stable or neutral bifurcation of path under quasistatic loading in the plastic domain is of great interest because, as for the elastic-plastic column, it may lead to maximum loads that are below the loads for instability of a perfect configuration. Not only is the result then more realistic, it may be easier to determine than instability of configuration. Satisfaction of the equations of continuing equilibrium is a clear as well as strong restriction. Hill [2], Rice [3] and many others following in their footsteps [4–7] have obtained important results on localization, or the formation of a shear band in a homogeneously stressed body with and without void formation, from this interesting extended Shanley point of view. Continuity of surface traction and traction rates across the boundary between the main body and the shear band, each deforming in its own homogeneous mode, gives rise to an eigenvalue problem for the bifurcation.

A high degree of stability for a workhardening material is ensured by an associated flow rule, normality of the plastic strain increment $\delta\varepsilon_{ij}^p$ or rate vector $\dot{\varepsilon}_{ij}^p$ to the current yield surface $f(\sigma_{ij}) = $ constant.

$$\dot{\varepsilon}_{ij}^p = \dot{\lambda}\,\partial f/\partial\sigma_{ij}, \tag{1}$$

$$\dot{\sigma}_{ij}\dot{\varepsilon}_{ij}^p > 0 \quad \text{for} \quad \dot{\varepsilon}_{ij}^p \not\equiv 0. \tag{2}$$

It, along with the further stabilizing elastic contribution,

$$\dot{\sigma}_{ij}\dot{\varepsilon}^e_{ij} > 0 \quad \text{for} \quad \dot{\sigma}_{ij} \not\equiv 0 \tag{3}$$

carries over to a high degree of stability for compact specimens [8]. Conversely, the instability in the small in the forward sense [9]

$$\dot{\sigma}_{ij}\dot{\varepsilon}_{ij} = \dot{\sigma}_{ij}(\dot{\varepsilon}^e_{ij} + \dot{\varepsilon}^p_{ij}) < 0 \tag{4}$$

that can result from a non-associated flow rule

$$\dot{\varepsilon}^p_{ij} = \dot{\Lambda}\,\partial g/\partial\sigma_{ij}, \quad g \not\equiv f \tag{5}$$

because $\dot{\sigma}_{ij}\dot{\varepsilon}^p_{ij} < 0$ for some range of $\dot{\sigma}_{ij}$ should be expected to lead to an early instability of configuration. Under triaxial test conditions, this instability will be seen to lead to limited but appreciable dynamic jumps in the configuration. For the simple models customarily employed, such instability in the conventional triaxial test would set in immediately upon entering the plastic range.

Stability of configuration means that the response with time to all possible infinitesimal (or perhaps merely small) perturbations remains infinitesimal (or small). Instability means that some possible infinitesimal perturbation will cause a finite change in geometry. Both the kinematics of the finite motion and the kinetics must be consistent with the postulated behavior of the material and the loading of the specimen.

2 The triaxial test

The usual triaxial test is shown in Fig. 1. A predetermined value of hydrostatic pressure, (axial) $\sigma_V = \sigma_H$ (radial) is applied. It is followed by an increase in the axial compressive stress σ_V alone. The figure is drawn in accord with the usual convention for geomaterials, which are unable to take much if any tension and generally exhibit high sensitivity to normal stress. Compressive, rather than tensile, stress is considered as positive.

At each stage of the increase of the axial loading, a relevant question is whether or not there is an alternate to the existing uniform state of stress σ^0_{ij} (σ_V, σ_H, σ_H) and deformation. The many localization or shear band studies reported predict that stable or neutral bifurcation of the

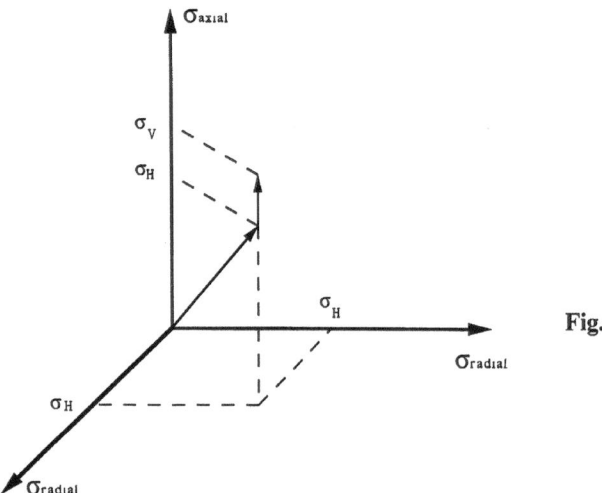

Fig. 1. The triaxial test

deformation pattern under increasing homogeneous stress in an elastic-plastic model ordinarily will not occur until the axial stress reached is at or near the peak of the axial stress vs. axial strain curve for the model in the absence of such bifurcation. These results certainly are in accord with intuition when the model obeys the normality relation for the plastic increment or rate of strain. A rising stress-strain curve indicates stability [8]. However, the prediction was found not to alter nearly as much as might be expected when the normality relation is not followed.

3 Stability in the small in the forward sense

The reason for expecting a quite different result for an elastic-plastic model with a non-associated plastic flow rule, shown in two dimensions in Fig. 2, is that a continuous loading path is likely to exist in stress space σ_{ij} that is unstable in the large as well as the small in the forward sense.

$$\text{(small)} \quad \dot{\sigma}_{ij} \dot{\varepsilon}_{ij} < 0, \tag{6}$$

$$\text{(large)} \quad (\sigma_{ij} - \sigma_{ij}^0) \, \dot{\varepsilon}_{ij} < 0. \tag{7}$$

Lack of normality at a smooth point on the current yield surface means that a range of stress paths must exist in the plastic domain for which, at each point, the rate of work done per unit of volume by the existing state of stress on the rate of plastic strain would exceed the energy dissipation rate.

$$\sigma_{ij}^0 \dot{\varepsilon}_{ij}^p > \sigma_{ij} \dot{\varepsilon}_{ij}^p \tag{8}$$

or

$$(\sigma_{ij} - \sigma_{ij}^0) \, \dot{\varepsilon}_{ij}^p < 0; \quad \int (\sigma_{ij} - \sigma_{ij}^0) \, d\varepsilon_{ij}^p < 0. \tag{9}$$

The alternate description, which avoids the use of the term "rate" for a time-independent material, is that the work done by the increment of stress on the increment of plastic strain is negative. If the contribution to the work by the elastic response is small compared with the plastic, the work done by the increment of stress on the total increment of strain, elastic plus plastic, would be negative.

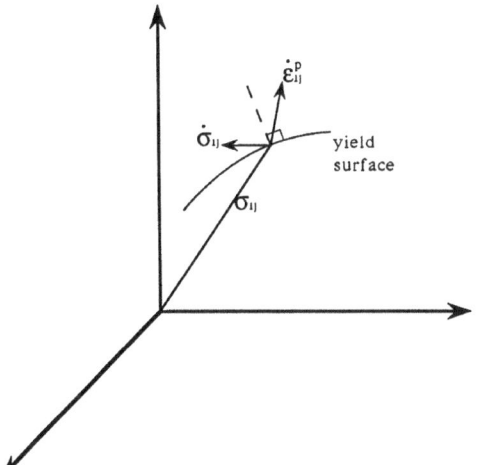

Fig. 2. A non-associated flow rule

Equivalently, the total work per unit volume done by the stresses of the existing state along the path of deformation then would exceed the energy stored plus the energy dissipated. The system would acquire kinetic energy. If the system is able to follow any such stress path, its initial static configuration is genuinely unstable. An infinitesimal disturbance will trigger off a finite accelerating response.

When the elastic response along the stress path is small compared with the plastic, such an unconstrained system is always unstable. A block situated at the saddle point of a tilted saddle-like frictionless surface provides a helpful but oversimple analog. Positive work must be done to move the block in most directions because the surface on which it rests slopes upward everywhere except for a narrow angular region on each side of the path of steepest descent. However, under vanishingly small disturbing forces continually applied in all possible directions, the block will slide in a generally downward direction in this angular region, picking up speed as it goes. Attachment of the block through a soft spring to the saddle point does not change the picture appreciably. The situation is also the same in the presence of friction with a sufficiently greater tilt of the surface to overcome the frictional resistance to sliding.

4 The Mohr-Coulomb criterion as an example

The genuine instability of configuration for a deformable body as a result of instability in the small in the forward sense is illustrated clearly with one of the simplest and most popular models for dry sand, an isotropic stress-hardening Mohr-Coulomb criterion of yield coupled with a non-associated flow rule. The geometric representation of each yield surface in principal stress

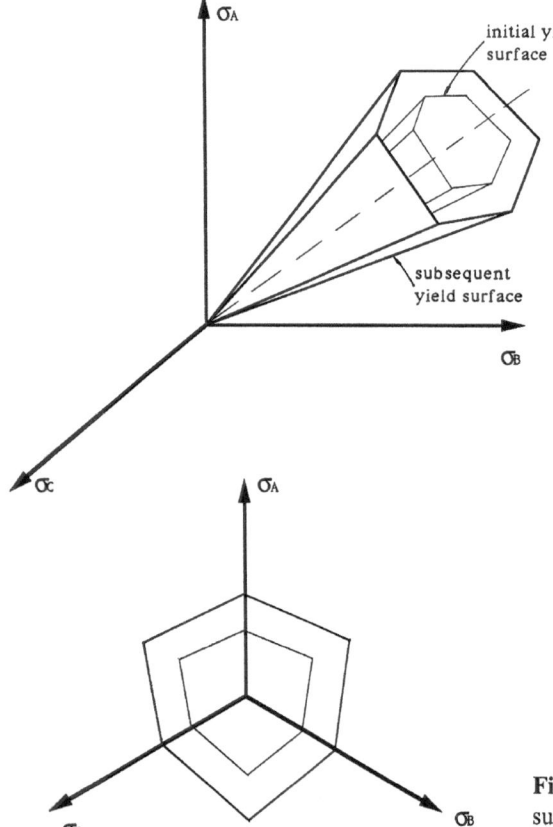

Fig. 3. Initial and subsequent Mohr-Coulomb yield surfaces

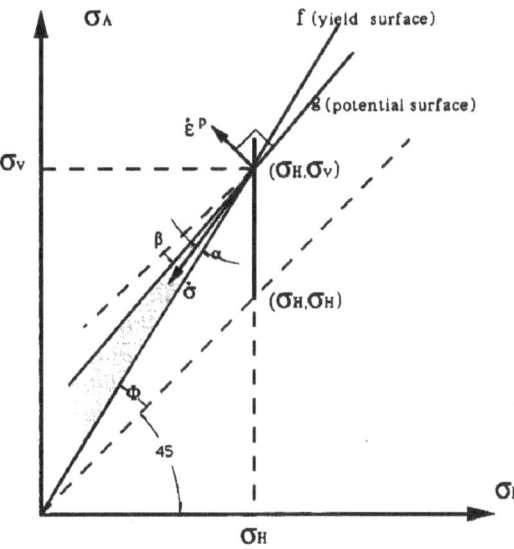

Fig. 4. Two dimensional stress space and the "wedge region"

space is a symmetric but not regular hexagonal pyramid with its apex at the origin and its axis equally inclined to the three principal stress axes. A potential function is selected that gives a plastic strain rate, or infinitesimal increment in plastic strain, involving much less volume change that would be required by normality to the Mohr-Coulomb surface.

As plastic deformation proceeds under any "outward" path of loading prior to failure, the Mohr-Coulomb yield surface simply expands its apex angles to become a larger pyramid with its apex remaining at the origin of principal stress space. Fig. 3 shows the initial and a subsequent yield surface in the space of the three principal stresses for the assumed isotropic hardening.

In the two dimensional principal stress space of Fig. 4, the radial line from the origin that is intersected by the rising stress path represents a plane in the three dimensional principal stress space that passes through the third principal stress axis. The normal to the plane therefore lies in the two dimensional plot and appears in full. Suppose each potential function line also represents a plane perpendicular to the plane of the paper. Its normal then also appears in full.

When axes of strain are superposed on the axes of stress, the infinitesimal plastic increment of strain associated with an infinitesimal increase in the axial stress in the plastic domain is represented by a vector in this space normal to the potential line. The plastic strain increment or rate vector makes an acute angle with the vector representing an increment in axial stress. Therefore the total work that would be done by such an increment of stress on the total increment of strain would be positive, because the elastic increment of strain similarly makes an acute angle with the increment of stress. Such stability in the small in the forward sense is stability in its customary meaning.

5 The wedge region

However, the result is very different when all permissible stress paths at the given point in stress space are examined. Consider the initial segment of a path of loading in the shaded "wedge region" between the current potential surface and the current yield surface, Fig. 4. The current yield surface is at a counterclockwise angle Φ to the 45 degree line; the current potential surface is at the counterclockwise angle β. The stress increment points outside the current yield surface of the model at an angle α, so that plastic deformation will occur and the plastic strain increment will be represented by a vector normal to the potential surface as before. Now the work done by the increment or rate of stress on the increment or rate of plastic strain will be negative; the angle between the two vectors is greater than a right angle by $\Phi - (\alpha + \beta)$.

$$\dot{\sigma}_{ij}\dot{\varepsilon}^p_{ij} = |\dot{\sigma}_{ij}| \, |\dot{\varepsilon}^p_{ij}| \cos\left[\pi/2 + \Phi - (\alpha + \beta)\right] < 0, \tag{10}$$

$$\cos\left[\pi/2 + \Phi - (\alpha + \beta)\right] = -\sin\left[\Phi - (\alpha + \beta)\right], \tag{11}$$

$$|\dot{\varepsilon}^p_{ij}| = (|\dot{\sigma}_{ij}| \sin \alpha)/h \tag{12}$$

where the absolute value signs denote the magnitude of the vectors and h is the plastic modulus.

The work done on the increment of elastic strain will still be positive. For isotropic elastic response, $\dot{\sigma}_{ij}\dot{\varepsilon}^e_{ij}$ is of order $|\dot{\sigma}_{ij}| \, |\dot{\varepsilon}^e_{ij}|$ with $|\dot{\varepsilon}^e_{ij}|$ of order $|\dot{\sigma}_{ij}|/E$ where E is the elastic modulus.

$$\dot{\sigma}_{ij}\dot{\varepsilon}^e_{ij} \quad \text{is of order} \quad |\dot{\sigma}_{ij}|^2/E. \tag{13}$$

Consequently it is possible that the work done by the increment of stress on the increment of total strain will remain positive and there will be stability in the small in the forward sense. However, if the positive contribution of the elastic response is smaller in magnitude than the negative one of the plastic, the path is unstable in the small. For the simplest and customary assumption of elastic response close to negligible in comparison with the plastic, instability in the small will be found for just about every wedge path direction.

When an associated flow rule is employed, $\beta = \Phi$ and $\cos\left[\pi/2 + \Phi - (\alpha + \beta)\right] = \sin \alpha$ is a positive rather than a negative quantity. Work then is not available for conversion to kinetic energy. It is clear also, when the elastic contribution is considered, than a very small deviation from normality is not sufficient to produce instability (a negative $\dot{\sigma}_{ij}\dot{\varepsilon}_{ij}$).

In the following sections, the instability resulting from an appreciable deviation from normality will be shown to be a genuine instability as ordinarily defined, just as for the block sliding on the inclined saddle surface. For the Mohr-Coulomb material model so often employed and related models, a wedge region exists at each and every stress point in the plastic domain. Therefore the configuration of a triaxial test specimen of any such model material would be unstable as soon as the plastic domain is entered. This is a logical consequence of using a particular type of model, not a statement of physical reality. It leaves open the very different and important problem of interpretation or choosing a useful model consistent with stable triaxial tests.

6 Comments on the requirements for a genuine instability

In order for a configuration of a material system under load to be unstable, it is necessary, but far from sufficient, for an alternate configuration to exist. Alternate configurations often exist, even for highly stable systems. A given configuration is unstable only if at least one possible infinitesimal (or perhaps small) set of disturbances will take the existing system along some continuous path to or beyond some finitely removed alternate configuration. In general the system will not follow an equilibrium path, but both its kinematics and kinetics must be permissible.

An axially loaded column provides a typical illustration. At a load well below the maximum load the column can carry, a strongly bent or buckled configuration is an alternate for the straight column under a given load. However, the barrier to such a change in configuration is very large. Modest perturbations of the load and geometry can not produce the transition from a straight to a strongly bent shape. On the other hand, a perfect axially loaded elastic column above its Euler buckling load will snap over to a strongly bent configuration with the smallest of perturbations of its geometry or axiality of loading. An elastic-plastic column below its maximum load carrying capacity needs a finite disturbance to bring it over to the falling branch of the load-deflection curve. The closer the load to the carrying capacity, the smaller the disturbance required. At the peak load, an infinitesimal perturbation makes the system run away.

7 An unstable stress path

The picture that is proposed for consideration is not the usual one of an equilibrium bifurcation pattern. Quite the contrary, although the pattern remains that of a homogeneously deforming inclined "shear" band, Fig. 5, of zero extensional strain rate [3] in the material plane instantaneously bounding the band and all parallel planes.

At any initial triaxial state of stress σ_V, σ_H in the plastic range, the state of principal stress in the band σ_A, σ_B, σ_C is chosen to follow a straight line wedge path from $\sigma_A = \sigma_V$, $\sigma_B = \sigma_H$ in the plane of two of the principal stresses, σ_A, σ_B, Fig. 4. This is a stress path of continuing plastic deformation in the A, B plane that continuously pushes the yield surface outward, producing the same surfaces for isotropic stress-hardening as postulated for the conventional triaxial test, Figs. 1 and 3. The controlling ratio of the shear stress to the normal stress on each critical plane perpendicular to the paper continuously increases although each of the principal stresses in the plane of the paper continuously decreases. The principal stress σ_C normal to the plane of the paper is the intermediate principal stress and does not affect the plastic deformation in this simple Mohr-Coulomb model. There is no plastic strain in the C direction so that $\dot{\varepsilon}_C = 0$ gives

$$\dot{\sigma}_C = \nu(\dot{\sigma}_A + \dot{\sigma}_B) \tag{14}$$

for an isotropic elastic response.

There is no difficulty in principle in following a curved instead of a straight wedge path and taking into account any changes along the path in material properties, including changes in the slope of the yield surface and the potential surface. For simplicity of calculation and description

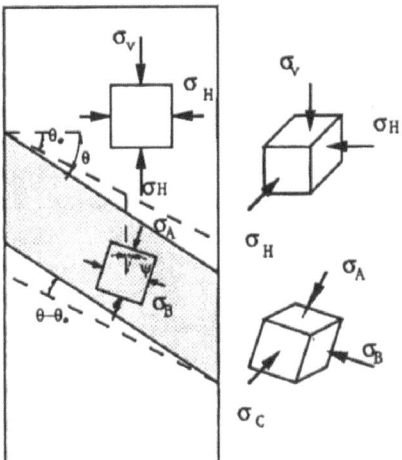

Fig. 5. Rotating shear band with rotated principal stresses

in most of what follows, however, the plastic modulus h and the direction of the plastic strain rate will be taken as constant and the change in the slope (angle Φ) of the current yield surface along the straight line wedge path will be ignored.

At each stage of the continuing plastic deformation the normal strain rate component $\dot{\varepsilon}_T$, as well as $\dot{\varepsilon}_C$, must be zero for the material instantaneously in the plane separating the deforming band from the non-deforming bulk of the material. This determines the angle of inclination θ of the band to the horizontal at each stage, an angle that, as will be seen, must continually increase as the wedge path is traversed.

8 A description of the simplest instability pattern

The simplest of all kinematic pictures for the instability demonstrates the need for rotation most clearly. For this purpose, the (changes in) elastic deformation will be neglected and the subsequent purely plastic plane strain deformation of the shear band will be taken as incompressible, $\beta = 0$. The strain rate $\dot{\varepsilon}_C$ is zero because the normal to the potential surface has a zero component in the intermediate principal stress direction. Incompressibility requires equal and opposite strain rate components in the other two principal stress directions, positive (compressive) in the direction A which initially is axial, and negative (extension) in the direction B which initially is horizontal. The plane of zero normal strain rate is at 45° to these principal directions. It divides the rest of the body from the band deforming in simple shear parallel to that 45° plane. The strain rate $\dot{\varepsilon}_N$ in the direction of the band thickness is zero. Therefore, if the band rotates, the axes of principal stress in the band must rotate by exactly the same amount to keep at 45° to the boundary and so maintain zero $\dot{\varepsilon}_T$.

For the shear band to develop and the system to accelerate or begin to run away, it is necessary that the kinetics agree with the kinematics as the state of stress changes with the deformation. Both the major and the minor principal stresses σ_A, σ_B decrease in magnitude as a wedge path is followed. A free body diagram of any upper portion of the body, Fig. 6, then will show a departure from the initial equilibrium state. There is a net force down along the boundary plane which would appropriately accelerate the system by a_T in the shearing direction T.

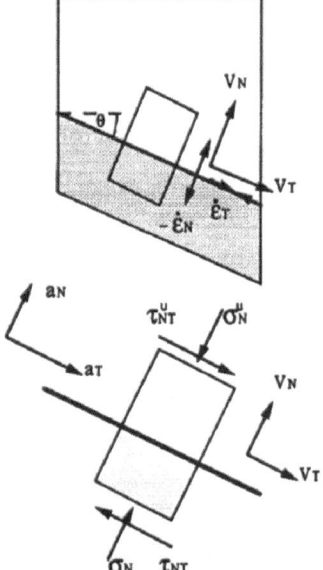

Fig. 6. Equivalent of a free body diagram

However, there is also a net compressive force directed into the plane (negative N direction) which would appear to improperly accelerate the upper portion by a_N in the direction of compression of the shear band. The tangential acceleration is proportional to the tangential force which in turn is proportional to the difference between the tangential components of stress τ_{NT}^u and τ_{NT} in the upper portion and in the band. The normal acceleration is proportional to the difference between the normal components of force which in turn is proportional to the difference between the normal components of stress σ_N^u, σ_N.

The picture can not be made kinetically consistent with a fixed orientation of the bounding plane. However, if the plane rotates toward the vertical as the wedge path is followed and the principal axes of stress in the shear band rotate with it, Fig. 5, remaining at $45°$ for this simple example, the rate of extensional strain in the dividing plane remains zero. An appropiate choice of the rate of rotation for each rate of principal stress change can produce the inward (negative a_N) acceleration needed. In general

$$a_T = dv_T/dt + \dot{\theta} v_N, \tag{15}$$

$$a_N = dv_N/dt - \dot{\theta} v_T \tag{16}$$

but in this simple picture, $\dot{\varepsilon}_N$ and therefore v_N is zero, the inward acceleration is a reflection of the clockwise rotation of a changing v_T.

9 The general instability pattern

In general, of course, the angle β will not be zero and the shear band will have a plastic strain rate component $\dot{\varepsilon}_N^p$ which ordinarily will be negative (an increase in volume). Therefore, v_N will be positive rather than zero.

For the system to be unstable, to pick up speed or kinetic energy, power must be fed into the system. The resultant force on the upper block, and therefore its acceleration, must make an acute angle with the resultant velocity.

$$\mathbf{F} \cdot \mathbf{v} > 0; \quad \mathbf{a} \cdot \mathbf{v} = a_N v_N + a_T v_T > 0. \tag{17}$$

Substitution for a_T, a_N in terms of the velocity exhibits the increase in kinetic energy as the instability accelerates from a zero start in its early stages.

$$v_N \, dv_N/dt + v_T \, dv_T/dt = d/dt[1/2(v_N^2 + v_T^2)] > 0. \tag{18}$$

Substitution in terms of the stress differences and strain rates gives:

$$(\sigma_N - \sigma_N^u)(-\dot{\varepsilon}_N) + (\tau_{NT}^u - \tau_{NT}) \dot{\gamma}_{NT} > 0 \tag{19}$$

where the first term is negative and the second, which provides the modest driving force for the limited instability, is positive.

This inequality is the equivalent of the earlier one for instability of the material of the band in the large in the forward sense, inequality (7).

$$(\tau_{NT} - \tau_{NT}^u) \dot{\gamma}_{NT} + (\sigma_N - \sigma_N^u) \dot{\varepsilon}_N < 0 \tag{20}$$

where the first term is negative because the shear strain rate component $\dot{\gamma}_{NT}$ is positive and opposite in the direction to the negative $\tau_{NT} - \tau_{NT}^u$. The second term is positive because the strain rate component $\dot{\varepsilon}_N$ is in the same direction as $\sigma_N - \sigma_N{}^u$. Both are negative in this sign convention.

As the angle θ increases, the exterior loading on the upper body causes a smaller and smaller (compressive) normal component of force proportional to $\sigma_N{}^u$ and shear force proportional to τ_{NT}^u. The forces on the plane bounding the shear zone also decrease as the stress components σ_A, σ_B in the shear zone decrease. However, the system rather soon runs out of driving force; $\tau_{NT}^u - \tau_{NT}$ itself goes to zero at a rather moderate value of θ beyond its initial value.

$$\tau_{NT}^u = [(\sigma_v - \sigma_H)/2] \sin 2\theta, \tag{21}$$

$$\tau_{NT} = [(\sigma_A - \sigma_B)/2] \sin 2(\theta - \psi). \tag{22}$$

Inequality (19) is no longer satisfield. Therefore the system will overshoot a little and then come to rest after a significant but not very large excursion of plastic strain.

Also, the elastic components of strain rate will be nonzero and must be taken into account when finding the plane of zero extensional strain rate. In contrast to the result for plastic strain rates or increments, rotation of a constant principal state of stress would produce an elastic strain rate in the material plane that instantaneously divides the shear zone from the upper portion of the body. Were it not for this contribution to the extensional strain rate, the linear relation between the changes in principal stress and the changes in elastic strain would again require the rotation of the principal stress directions in the band to follow exactly the rotation of the band $\theta - \theta_0$ for a straight line wedge path. When the elastic changes are small compared to the plastic, the deviation of ψ from $\theta - \theta_0$ will not be great, Fig. 5.

Although it might be comforting to obtain a complete dynamic solution to this complicated problem, the instability is clear without doing so. Also, the details will be very different for each wedge path, for each size of specimen and width of shear band, and for each magnitude of elastic vs. plastic response. All that is needed is to be sure that a solution is possible. The absence of any intrinsic time scale for elastic-plastic (time-independent) material provides great flexibility.

Clearly, the larger the elastic strain increments compared with the plastic, the more restricted the paths in the wedge region that are unstable. When the elastic response overwhelms the plastic, all paths are stable. It also is apparent that the gradual addition of elastic response to a purely plastic one can produce at most a gradual change in the onset of instability and in the overall response.

10 Concluding remarks

A calculated bifurcation or instability is always of interest. Its practical significance, however, depends both upon when it sets in and upon the immediate or subsequent extent of the departure of the system from the initial configuration. Unfortunately, it is seldom possible to be certain that the lowest instability or bifurcation load of a complex irreversible system has been found. Only rarely, as in the standard triaxial test of a conventional Mohr-Coulomb model, is the configuration unstable for every state of stress in the plastic range.

Instability of configuration is much more dramatic than instability of path. In either case, however, it is necessary to examine the behavior of the system beyond its initial unstable response in order to assess the importance of the instability. It is very difficult to follow a shear band instability in full detail. However, it is possible to estimate the extent of the change in strain or displacement that is likely to occur. An unstable shear band that develops at a given load and causes catastrophic failure of the specimen is quite different from the moderate one described here which ceases activity and produces a limited total displacement. Both are far more troublesome than a shear band which does not deform at all without a further increase in load.

Acknowledgements

The authors wish to thank the Office of Naval Research, Solid Mechanics Program, Mechanics Division, Dr. R. Barsoum, for the support of this work under Grant Number N00014-87-J-1193. It gives us great pleasure to dedicate this paper to Professor Bruno Boley, a genuinely stable person.

References

[1] Shanley, F. R.: Inelastic column theory. Journal of the Aeronautical Sciences 14, 261 – 268 (1947).

[2] Hill, R.: Acceleration waves in solids. J. Mech. Phys. Solids 10, 1 – 16 (1962).

[3] Rice, J. R.: The localization of plastic deformation. In: Theoretical and Applied Mechanics (Koiter, W. T., ed.), pp. 207 – 220, Proceedings of the 14th International Congress of Theoretical and Applied Mechanics, North-Holland 1976.

[4] Rudnicki, J. W., Rice, J. R.: Conditions for the localization of deformation in pressure-sensitive dilatant materials. J. Mech. Phys. Solids 23, 371 – 394 (1975).

[5] Needleman, A.: Non-normality and bifurcation in plane strain tension and compression. J. Mech. Phys. Solids 27, 231 – 254 (1979).

[6] Mear, M. E., Hutchinson, J. W.: Influence of yield surface curvature on flow localization in dilatant plasticity. Mechanics of Materials 4, 395 – 407 (1985).

[7] Loret, B., Harireche, O.: Acceleration waves, flutter instabilities and stationary discontinuities in inelastic porous media. J. Mech. Phys. Solids 39, 569 – 606 (1991).

[8] Drucker, D. C.: Plasticity. In: Structural Mechanics (Goodier, J. N., Hoff, N. J., eds.), pp. 407 – 455, Proceedings of the 1st Symposium on Naval Structural Mechanics 1958, Pergamon Press 1960.

[9] Drucker, D. C.: On the postulate of stability of material in the mechanics of continua. J. de Mécanique 3, 235 – 249 (1964).

Authors' address: Prof. D. C. Drucker, Department of Aerospace Engineering, Mechanics and Engineering Science, University of Florida, Gainesville, FL 32611-2031, U.S.A.

Acta Mechanica (1992) [Suppl] 3: 173−179

On creep buckling of structures

T. H. Lin, Los Angeles, U.S.A.

Summary. Criteria of creep buckling of structures are discussed. The useful life of the structures is controlled by the tolerable deformation, which should be the criterion for design. Current creep strain affects creep strain rate as indicated by the mechanical equation of state. Past creep strain may change the time-independent stress-strain, relation, which determines the elastic-plastic buckling of a number of structures. The different effects of creep strain and initial strain on the deformation rate of the structure is shown.

1 Introduction

Two different criteria have been used in the treatment of buckling of structures subject to creep. That of Robotnov and Shesterikov, 1957 [1] consists of examining the structure stability in the dynamic sense. The equations of motion of the structure are linearized for small displacements in the region of some state through which the structure passes and the structure is considered stable in that state if a decaying response is indicated and unstable if an increasing response occurs. The other approach which is mainly associated with Hoff and his various co-workers [2 − 6] is based on the use of quasi-static equations to describe the displacement of the creep structure. Creep buckling is considered to occur if these equations indicate an infinite displacement rate or infinite displacement [4, 5], Hoff has indicated that this creep buckling is essential in assessing the useful life of a structure.

2 Criterion of maximum tolerable deformation

In engineering applications, we are interested in the useful life of the structure, which is governed by the time to reach the tolerable deformation. This deformation can be developed by creep, the time-dependent strain or by the plastic, the time-independent strain or a combination of these two. If the structure has an infinite displacement rate, it is possible after the structure develops certain displacement, the displacement rate may become finite. Then the life of the structure still is governed by the maximum tolerable deformation.

3 Stress-strain-time relationship

When an indeterminate structure is loaded at an elevated temperature, creep strain occurs. Creep strain gives redistribution of stress, which generally reduces the peak stress in the structure. The stress at a point in the structure varies with time. Creep tests generally were done under constant load or constant stress. To apply these test data to creep analysis of structures with stresses varying with time, one commonly used assumption is the existence of "Mechanical equation of state" between creep rate, stress, temperature and the current creep strain (Lin, 1968 [7], Johnson, 1949 [8]). For a uniaxial loading, this equation is written as

$$\dot{e}_c = F(\sigma, T, e_c) \tag{1}$$

where F denotes a function e_c is the uniaxial creep strain, σ the stress T is the temperature and dot denotes the time rate. Under constant temperature, Eq. (1) reduces to

$$\dot{e}_c = F(\sigma, e_c). \tag{2}$$

If the structure is subject to a multiaxial stress τ_{ij}. We let the effective strain be defined as

$$\dot{e} = \frac{\sqrt{2}}{3} [(e_{11} - e_{22})^2 + (e_{22} - e_{33})^2 + (e_{33} - e_{11})^2 + 6(e_{12}^2 + e_{23}^3 + e_{31}^2)]^{1/2} \tag{3}$$

and the effective stress as

$$\sigma^* = \frac{1}{\sqrt{2}} [(\tau_{11} - \tau_{22})^2 + (\tau_{22} - \tau_{33})^2 + (\tau_{33} - \tau_{11})^2 + 6(\tau_{12}^2 + \tau_{23}^3 + \tau_{31}^2)]^{1/2}. \tag{4}$$

Then the mechanical equation of state for the uniaxial loading is generalized to one for multiaxial loading

$$\dot{e}_c^* = F(\sigma^*, T, e_c^*). \tag{5}$$

It is further assumed that the principal axes of strain rate coincide with those of principal stresses.

$$\frac{\dot{e}_{11_c}}{S_{11}} = \frac{\dot{e}_{22_c}}{S_2} = \frac{\dot{e}_{33_c}}{S_{33}} = \frac{\dot{e}_{12_c}}{S_{12}} = \frac{\dot{e}_{23_c}}{S_{23}} = \frac{\dot{e}_{31_c}}{S_{31}} = \frac{\dot{e}_c^*}{2\sigma^*} \tag{6}$$

where δ_{ij} are the deviatoric stress components, Eqs. (5) and (6) are used to calculate the incremental creep strain components.

Dorn, 1954 [9] has shown the dependency of creep on temperature as

$$e_c = \phi \left(t\varepsilon^{-\frac{\Delta H}{RT}} \right) \tag{7}$$

where ϕ denotes a function, ΔH is the activation energy, R is the gas constant, T is in degrees Kelvin, and ε is the base of the natural logarithm. The activation energy for creep could be determined from two creep tests conducted under the same stress at two different temperatures. If the times to reach the same strain at two temperatures T_1 and T_2 are t_1 and t_2 respectively, we have

$$t_1 \varepsilon^{-\frac{\Delta H}{RT_1}} = t_2 \varepsilon^{-\frac{\Delta H}{RT_2}} \tag{8}$$

and ΔH can be found. In the range of $T = 422\,°K$ to $525\,°K$, the above relation represents the experimental data quite well. If the structure is loaded under constant temperature, Eq. 5 reduces to

$$\dot{e}_c^* = F(\sigma^*, \dot{e}_c^*). \tag{9}$$

This gives a stress-strain-time relation under constant temperature. For example, this relation of 75S-T6 aluminium alloy at 600 °F in uniaxial loading has been given by Higgins, 1952 [10], as

$$e_c = A\varepsilon^{B\sigma}t^K \tag{10}$$

where $A = 2.64 \times 10^{-7}$, $B = 1.92 \times 10^{-3}$, $K = 0.66$, σ is in p.s.i., t is in hrs.

From (8)

$$t = \left(\frac{e_c}{A\varepsilon^{B\sigma}}\right)^{\frac{1}{K}}$$

$$\dot{e}_c = K e_c^{\frac{K-1}{K}}(A\varepsilon^{B\sigma})^{\frac{1}{K}}. \tag{11}$$

For multiaxial loading, this equation is written as

$$\dot{e}_c = K(\dot{e}_c^*)^{\frac{K-1}{K}}\left(A\varepsilon^{B\sigma^*}\right)^{\frac{1}{K}}. \tag{12}$$

Eqs. (9) and (12) satisfy the mechanical equation of state.

4 Stresses caused by creep and/or plastic strain

In indeterminate structures, creep and plastic strains cause a redistribution of stress which generally decreases the peak stress. Creep rate depends on the stress after redistribution, so this stress needs to be calculated. To calculate these stresses, we imagine that the structure is unloaded. The creep strain and/or the plastic strain remains and causes a residual stress field. This residual stress can be calculated using elastic solutions by the analogy between the creep and/or plastic strain and the applied forces in an elastic solid as shown ny Lin, 1968 [7].

Let τ_{ij} denote the stress components and e_{ij}, the strain components. Let the superscript "c" denote creep, and "p" denote plastic and double prime denote the sum of creep and plastic strains

$$e''_{ij} = e^c_{ij} + e^p_{ij}. \tag{13}$$

For an solid of isotropic elasticity, we have

$$\tau_{ij} = \delta_{ij}\lambda(e_{kk} = e''_{kk}) + 2\mu(e_{ij} - e''_{ij}) \tag{14}$$

where λ and v are Lame's constants, the repetition of subscript denotes summation from 1 to 3 and δ_{ij} is the Kronecker delta. The condition of equilibrium is given as

$$\tau_{ij,j} + F_i = 0 \tag{15}$$

in the interior of the body where subscript j after comma denotes differentiation with respect to j-axis and F_i is the body force per unit volume along x_i-direction. Equilibrium conditions at the surface Γ gives

$$S_i = \tau_{ij}v_j \tag{16}$$

where S_i is the traction force per unit area along x_i-direction and v_j is the cosine of the angle between the normal v to the surface and x_j-axis. Substituting (14) into (15) and (16) yields

$$\lambda\delta_{ij}e_{kk,j} + 2\mu e_{ij,j} + (-\lambda\delta_{ij}e''_{kk,j}) - F_i = 0 \tag{17}$$

$$S_i + (\lambda\delta_{ij}v_je''_{kk} + 2\mu v_je''_{ij}) = \lambda\delta_{ij}v_je_{kk} + 2\mu v_je_{ij}. \tag{18}$$

Equation (17) gives partial differential equations with e_{ij}'s as the dependent variables and x_1, x_2, x_3 as the independent variables. Equation (18) gives the boundary conditions on the solid. It is seen that the terms in the parentheses in the above two equations are equivalent respectively to the body and surface forces in causing e_{ij}. These forces are called by Lin, 1968 [7], as the equivalent body force and surface forces. This analogy between creep strain and applied force was first used by Lin, 1962 [11] in the analysis of plate bending with strain-hardening creep. If we replace the creep strain by thermal strain, Eqs. (17) and (18) give Duhammel's analogy for thermal stresses.

With a given creep and/or plastic strain distribution e_{ij}'', the equivalent body force F_i and the surface force S_i are calculated. Then the strain distribution e_{ij} is found from elastic solutions. From Eq. 14, the residual stress yield τ_{ij}^r is readily calculated. Then the stress field in this body is

$$\tau_{ij} = \tau_{ij}^r + \tau_{ij}^a \tag{19}$$

where τ_{ij}^a is the stress field caused by the applied load.

5 Initial strain vs. creep strain

The similarly of creep strain and initial strain on creep buckling has been discussed by Hayman, 1978 [12, 13]. Initial strain e_{ij}^0 can be considered as one type of inelastic strain e_{ij}''. If we replace e_{ij}'' by e_{ij}^0, we will obtain the same relations given by Eqs. (17) and (18). This means if we have an initial strain field e_{ij}^0 identical to the creep strain e_{ij}^c, the strain e_{ij} and stress τ_{ij} fields would be identical.

However from the mechanical equation of state, the creep rate depends on both the stress and the creep strain. In the case of initial strain, the creep strain is zero while in the case of creep strain, it is not. Hence the creep rate in the case of initial strain is different from what with creep strain. This is illustrated in the following simple example.

A column of a uniform cross-section of a depth h of 1/4 inch, a width b of 5/8 inch and a length l, of $8''$ is simply supported at two ends. The initial curvature is a single half sine curve with an amplitude of 0.002 in. The column is of aluminium alloy 75ST-6 loaded at 600 °F. The stress-strain-time relationship of this material is given in Eqs. 10 and 11.

Referring to Fig. 1, with w_0 equal to the initial deflection, w_1, the additional deflection, and P, the compressive load. The elastic strain is the difference between total strain and creep strain

$$\sigma = E(e - e_c) : \Delta\sigma = E(\Delta e - \Delta e_c). \tag{20}$$

From the condition that plane sections before loading remain plane during loading

$$e = e_0 + k\eta, \quad \Delta e = \Delta e_0 + \eta\Delta k \tag{21}$$

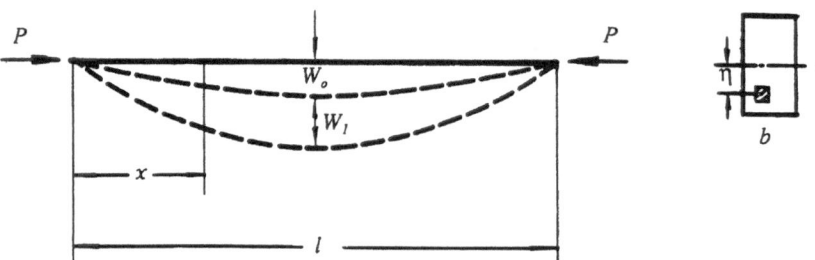

Fig. 1. A column with creep

where e_0 is the strain at the geometrical centroidal axis, k is curvature and η the distance below the centroidal axis

$$\Delta P = \int \Delta \sigma \, dA = E \int (\Delta e_0 + \eta \Delta k - \Delta e_c) \, dA. \quad \text{with} \quad \Delta P = 0,$$

$$\Delta e_0 = \frac{\int \Delta e_c \, dA}{A} = \bar{e}_c$$

(22)

$$\Delta M = \int \Delta \sigma \eta \, dA = E \int [\Delta e_0 + \eta \Delta k - \Delta e_c] \eta \, dA = EI\Delta k - E \int \Delta e_c \eta \, dA \tag{23}$$

where I is the moment of inertia of the cross-section. Referring to Fig. 1,

$$\Delta M = -P\Delta w \tag{24}$$

Equating Eq. (23) to Eq. (24) and considering small deflection with

$$\Delta k \cong \frac{d^2 \Delta w}{dx^2},$$

we obtain the equation of equilibrium

$$\frac{d^2 \Delta w}{dx^2} + \frac{P}{EI} \Delta w = -\frac{P}{EI} \left(-\frac{E}{P} \int \Delta e_c \eta \, dA \right). \tag{25}$$

The parenthesis term has the same effect as the initial deflection w_0 in causing the deflection Δw as shown by Lin 1968 [7]. Expressing this parenthesis term as a sine series.

$$\left(-\frac{E}{P} \int \Delta e_c \eta \, dA \right) = \Sigma \Delta A_n \sin \frac{n\pi k}{l}$$

$$\Delta w = \Sigma \frac{\Delta A_n}{\dfrac{P_n}{P} - 1} \sin \frac{n\pi k}{l} = \Sigma \Delta B_n \sin \frac{n\pi k}{l}$$

(26)

where $P_n = n^2 \pi^2 \dfrac{EI}{l^2}$.

The Young's modulus E at this temperature is 5.2×10^6 p.s.i. The critical compressive stress

$$\sigma_{cr} = \frac{P_{cr}}{bh} = 4\,170 \text{ p.s.i.}$$

The applied compressive stress $\dfrac{P}{A}$ is taken to be $2\,500$ p.s.i. The initial deflection is taken as a single sine half curve with an amplitude of $0.002''$. Under loading, this deflection is magnified to

$$w = \frac{.002}{1 - 2\,500/4\,170} = 0.005''.$$

The stress at different points in the column

$$\sigma = \frac{P}{A} \pm \frac{(Pw)\eta}{A} = \frac{P}{A} \left(1 \pm \frac{w\eta}{\varrho^2} \right) \tag{27}$$

where ϱ is the radius of gyration

$$\varrho^2 = \frac{I}{A} = \frac{1}{192} \text{ in}^2.$$

The deflection under load P at the initial instant is

$$w = .005 \sin \frac{\pi k}{l}.$$

The creep strain is calculated at five sections $x = \frac{l}{6}, \frac{l}{4}, \frac{l}{2}$. This strain is symmetrical to the mid-section of the column. At each section the stress σ is calculated at $\eta = \pm\frac{h}{2}, \pm\frac{3}{5}h, \pm\frac{h}{2}, \pm\frac{h}{8}$ and 0. From these stresses, the creep strains e_c at $t = \frac{1}{2}$ hour were calculated at these points. From these e_c's the values of $\left(-\frac{P}{E} \int e_c \eta \, dA \right)$ were calculated at the different sections. Then the coefficients A_n and B_n of the sine series, were calculated from Eq. 26. Δe_0 was found from (22). Then

$$\Delta k = \Sigma - \frac{n^2\pi^2}{l^2} \Delta B_n \sin \frac{n\pi k}{l} \tag{28}$$

$$\Delta e = \Delta e_0 + \eta \Delta k \tag{29}$$

$$\Delta\sigma = E(\Delta e - \Delta e_c). \tag{30}$$

Adding this incremental stress to the stress at the start of this time increment, we obtain this stress at the beginning of the next time interval. This process is repeated for the next time increment. The first two time increments were taken to be $\frac{1}{2}$ hr. and the subsequent increments were 1 hr.

The creep strains developed in this 4th and the subsequent time increment were calculated for two cases; one with creep strain taken from the end of the 3rd time increment and one with zero creep strain. The latter corresponds to the column with an initial strain distribution same as the

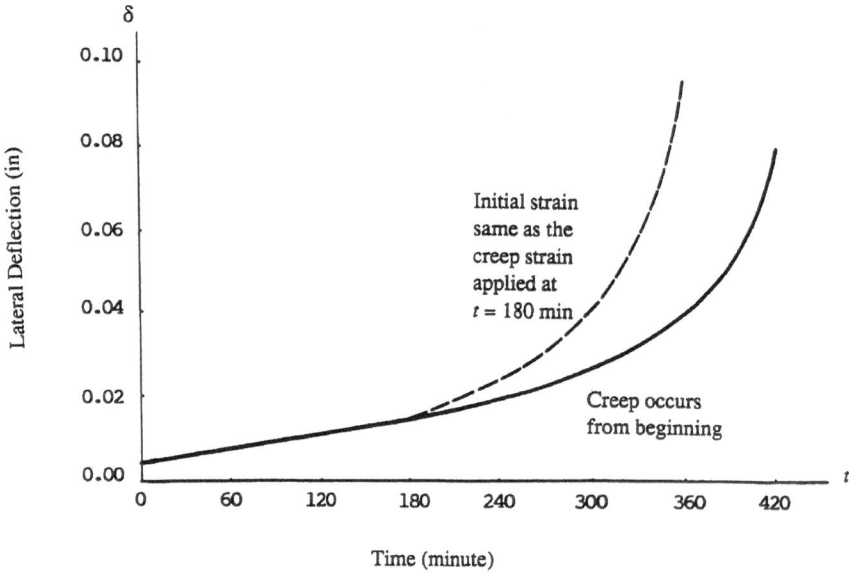

Fig. 2. Lateral deflection versus time

creep strain from previous creep. The calculated deflection at the mid-section of this column for the two cases are shown in Fig. 2. It is seen that the case with an initial strain as the creep strain, creeps more in the 4th and subsequent time increments.

If at the end of the 4th time increment, the load P in increased. If this increase of load is a function of time, we can calculate the creep deformation of this column versus time by the same method as described. However if the load is instantaneously increased, so the incremental creep strain is negligible. The early creep strain likely changes the time-independent stress-strain curve of the material. Then this column can be treated as an elastic-plastic column. This detail of this analysis was shown by Lin, 1949 [14].

6 Conclusions

The useful life of a structure loaded at an elevated temperature is generally limited by the maximum allowable deformation. This should be taken as the criterion for creep design. To calculate this deformation, we need to have a stress-strain-time relationship of the material for varying stress. To take care of this, the mechanical equation of state is commonly used. To calculate the stress field after the occurrence of creep, analogy between the creep strain and the applied load in causing strain and stress fields in an isotropic elastic body is shown. This seems to give a direct approach to evaluate the useful life of a structure at an elevated temperature.

References

[1] Robotnov, N., Yu, Shesterikov, S. A.: Creep stability of columns and plates. J. Mech. and Phys. of Solids vol. 6, No. 1, (1956).
[2] Hoff, N. J.: Buckling and stability. J. Roy. Aeronaut. Soc., Jan. (1954).
[3] Hoff. N. J.: A survey of the theories of creep buckling. Proc. 3rd S. Nat. Congr. of Appl. Mech., Providence, R. I. (1958).
[4] Hoff. N. J.: Axially symmetric creep buckling of circular shells in axial compression. J. Appl. Mech. **35**/3 (1968).
[5] Hoff. N. J.: On the transition from axisymmetric to multilobed creep buckling. Theory of Aircraft Structures, Delft (1972).
[6] Levi, I., and Hoff, N. J.: Interaction between axisymmetric and non-symmetric creep buckling. Proc. IUTAM Symp., Gothenburg Creep in Structures, p. 405−422 (1970).
[7] Lin, T. H.: Theory of inelastic structures. pp. 43−50, 65−68, 175−177, New York: Wiley 1968.
[8] Johnson, A. E.: Creep and relaxation of metals at high temperatures. Engineering **168**, 237 (1949).
[9] Dorn, J. E.: Some fundamental experiments on high temperature creeps. J. Mech. Phys. Solids **3**, 85−116 (1954).
[10] Higgins, T. P., Jr.: Effect of creep on column deflection. Preprint No. 385, Inst. Aero. Sc. 1952.
[11] Lin, T. H.: Bending of a plate with nonlinear strain-hardening creep. pp. 215−228, Proc. IUTAM Symp: Creep in Structures 1962.
[12] Hayman, B.: Aspects of creep buckling I. The influence of post buckling characteristics. Proc. Roy. Soc. (London) **364**, 394−414 (1978).
[13] Hayman, B.: Aspects of creep buckling II. The effects of small deflection approximations on predicted behavior. Proc. Roy. Soc. (London) **A 364**, 415−433 (1978).
[14] Lin, T. H.: Inelastic column buckling. J. Aerospace Sci. **14**, 159−172 (1950).

Author's address: Prof. T. H. Lin, Department of Civil Engineering, University of California, Los Angeles, CA 90024-1593, U.S.A.

Acta Mechanica (1992) [Suppl] 3: 181–190

A lower bound on bifurcation buckling
of viscoplastic structures

S. R. Bodner, Haifa, Israel

Summary. The buckling of structures of elastic-viscoplastic materials is a stability problem that does not admit a realistic bifurcation formulation in the classical manner. In the absence of imperfections and inertial effects, the standard bifurcation criterion leads only to elastic buckling since an instantaneous jump in strain rate would develop at the critical condition. However, an expression for the "short time" inelastic tangent modulus at the pre-buckling strain rate can be developed from an appropriate incremental constitutive theory and this can be used in the quasi-static bifurcation buckling condition appropriate to the structure and loading. Such a buckling value can be interpreted as a lower bound on the actual instability condition. For the case of structures with initial imperfections, the calculation of local inelastic tangent moduli at the current state should lead to close correspondence between bifurcation and instability. Under creep conditions, the procedure gives approximate creep buckling times for both perfect and initially imperfect structures. For situations where the buckling mode generates abrupt changes in the multiaxial stress state, modifications to the reference constitutive theory are required to properly represent the governing physics. In this manner, the procedure seems capable of indicating buckling values consistent with test results without relying on a "deformation" type plasticity theory.

1 Introduction

The continued study of buckling of structures in the inelastic range has raised a number of fundamental questions many of which have been resolved for the condition of rate independent plasticity. Consideration of the more physically realistic time dependence of inelastic deformation has added to the difficulties and apparent paradoxes. For example, it has been shown by Tvegaard [1] and others that bifurcation of a perfect elastic-viscoplastic column can occur only at the elastic buckling load according to the direct application of Hill's criterion. This is due to the initial elastic response of an elastic-viscoplastic material to an instantaneous change in strain rate, which is indicated by all relevant constitutive equations. The active tangent modulus for the instantaneous change in strain rate supposed in the bifurcation analysis is therefore the elastic value. However, elastic buckling is not realized in accurately constructed structures that are predicted by the standard (rate independent) bifurcation theory to buckle in the inelastic range, even though the materials exhibit some strain rate sensitivity. As a consequence of this interesting but unrealistic result, the examination of viscoplastic column buckling is treated as the instability condition of the load-deformation behavior of an initially imperfect structure, e.g. Tvegaard [2].

An alternative approach to the perfect viscoplastic column problem is to utilize the "short time" tangent modulus of the material at the pre-buckling strain rates in the governing bifurcation criterion. This modulus could be obtained by means of a suitable constitutive theory. Critical loads calculated in this manner can be interpreted as lower bounds on the buckling condition. A bifurcation criterion that does not need information on initial imperfections nor requires an associated deformation analysis should be practically useful even if it applies only in an approximate manner to the actual situation. For loadings leading to creep deformations, the constitutive theory should be capable of indicating the overall response of the structure as well as the short time response to an incremental change in loading which is the quantity used in the criterion.

In the case of bifurcation modes that generate new stress components, the reference constitutive theory has to properly represent the appropriate response characteristics of the event. Most simple viscoplastic theories would require some modifications to do so. These refinements would be motivated by observatins on "corner-turning" tests and are a rough analogy to the concept of generation of a corner in yield surface theories.

The procedure for calculating the short time tangent modulus by means of the unified constitutive theory of the author and Partom (B-P) and associates was given in an earlier paper by Bodner, Naveh and Merzer [3]. The present paper presents an expanded discussion of the method and describes its application to creep buckling problems. Treatment of creep buckling has had its own history of controversial development and the relation of the proposed procedure to more conventional methods is discussed.

Regarding bifurcation modes that generate new stress components, the short time, lower bound tangent moduli indicated by the reference constitutive theory would correspond to bifurcation loads that are unrealistically low. On the other hand, conventional plastic flow theory with a smooth yield surface and the associated normality condition leads to unrealistically high bifurcation loads such as the elastic buckling value in some cases. It seems that appropriate refinement of the constitutive theory guided by results of tests on basic material response characteristics can serve to obtain reasonable buckling load predictions.

2 Determination of the plastic tangent modulus

The reference small strain, elastic-viscoplastic constitutive model (B-P) has been recently reviewed by Bodner [4] and the relevant equations are summarized in Table 1. As noted, it is a unified theory in the sense that all inelastic deformations are included in the plastic strain rate variable $\dot{\varepsilon}_{ij}^p$. In this context, plastic flow at various strain rates, creep, and stress relaxation are response features corresponding to different loading and governed by the same set of equations. The formulation does not require a yield criterion or loading/unloading conditions so that both elastic and plastic strain rate components, Eq. (1), are generally non-zero. Inelastic deformations are determined by the flow law, Eq. (2), and by the kinetic equation (3) which relates the invariant of plastic strain rate $D_2{}^p$ to the stress invariant J_2. The constant D_0 in Eq. (3) would be the limiting plastic strain rate in shear and n controls strain rate sensitivity and influences the level of the flow stress. Isotropic and directional hardening variables, Z^I and Z^D, which appear combined in the kinetic equation as Z, Eq. (5), are load history dependent according to Eqs. (6), (7) and (8). These evolution equations contain terms indicating increases of hardening due to plastic work and negative terms for thermal (static) recovery of hardening. Various material constants in the full set of equations would be functions of temperature, especially n and those in the thermal recovery terms.

For buckling problems, the material property of interest is the total tangent modulus E_T which, for proportional loadings, can be simply obtained from the plastic tangent modulus

Table 1. A summary of the elastic-viscoplastic constitutive model

---

Decomposition of strain rate:
$$\dot{\varepsilon}_{ij} = \dot{\varepsilon}^e_{ij} + \dot{\varepsilon}^p_{ij} + \alpha \dot{T} \delta_{ij} \tag{1}$$

Flow law:
$$\dot{\varepsilon}^p_{ij} = \lambda s_{ij}; \quad \dot{\varepsilon}^p_{kk} = 0 \tag{2}$$
with $s_{ij} = \sigma_{ij} - (1/3)\,\delta_{ij}\sigma_{kk}$

Kinetic equation:
$$D_2{}^p = D_0{}^2 \exp\left[-\left(\frac{Z^2}{3J_2} \right)^n \right] \tag{3}$$
with $D_2{}^p = (1/2)\,\dot{\varepsilon}^p_{ij}\dot{\varepsilon}^p_{ij}$; $J_2 = (1/2)\,s_{ij}s_{ij}$

from (2):
$$\lambda^2 = D_2{}^p / J_2 \tag{4}$$

Total hardening variable,
$$Z = Z^I + Z^D \tag{5}$$

Evolution equations of internal variables:

a. Isotropic hardening, Z^I:
$$\dot{Z}^I = m_1[Z_1 - Z^I]\,\dot{W}_p - A_1 Z_1 \left[\frac{Z^I - Z_2}{Z_1} \right]^{r_1} \tag{6}$$
with $Z^I(0) = Z_0$; $\dot{W}_p = \sigma_{ij}\dot{\varepsilon}^p_{ij} = (3J_2)^{1/2} \cdot \left(\frac{4}{3} D_2{}^p \right)^{1/2} = \sigma_{\text{eff}}\dot{\varepsilon}^p_{\text{eff}}$; $W_p(0) = 0$

b. Directional hardening, β_{ij}, Z^D:
$$\dot{\beta}_{ij} = m_2(Z_3 u_{ij} - \beta_{ij})\,\dot{W}_p - A_2 Z_1 \left[\frac{(\beta_{kl}\beta_{kl})^{1/2}}{Z_1} \right]^{r_2} v_{ij} \tag{7}$$
with $\beta_{ij}(0) = 0$; $u_{ij} = \sigma_{ij}/(\sigma_{kl}\sigma_{kl})^{1/2}$, $v_{ij} = \beta_{ij}/(\beta_{kl}\beta_{kl})^{1/2}$;

and the scalar measure of directional hardening is defined by:
$$Z^D = \beta_{ij}u_{ij} \tag{8}$$

Material constants:

D_0, n, Z_0, Z_1, Z_2, Z_3, m_1, m_2, A_1, A_2, r_1, r_2, α, and elastic constants; with $n = n(T)$
(In most cases can set $Z_2 = Z_0$, $A_1 = A_2$, $r_1 = r_2$)

$E_T{}^p$ and the elastic value E. For non-proportional loadings, the various stress and total and plastic strain rate components at the current state have to be considered in the basic definition of E_T as the ratio of the current effective stress rate to the current effective strain rate. Alternatively, the angle between the total strain rate and the plastic strain rate, in addition to E and $E_T{}^p$, will enter the calculation for the total tangent modulus. It is noted that it is inappropriate to deduce the applicable plastic tangent modulus from the overall response behavior of a structure to a particular loading condition. In fact, it is the material response to an increment of stress in the buckling mode superimposed on the overall response that is sought. Assuming that the constitutive model used for the structural analysis is properly representative of the material behavior, a direct calculation of the plasic tangent modulus from the constitutive equations could be performed. That should be obtainable from the evolution equations for hardening and the basic form of the kinetic equation in terms of state quantities such as the current values of stress and the hardening variables. As such, $E_T{}^p$ would not depend directly on the structural response (deformation) parameters. The exercise for determining $E_T{}^p$ was reported in [3] and is outlined here.

By means of the kinetic equation, (3), the plastic tangent modulus can be expressed as:

$$E_T{}^p = \frac{d\sigma_{\text{eff}}}{d\varepsilon^p_{\text{eff}}} = \frac{d}{d\varepsilon^p_{\text{eff}}}\left[Z \cdot f(D_2{}^p / D_0{}^2) \right] \tag{9}$$

where $\sigma_{\text{eff}} = \sqrt{3J_2}$, $\dot{\varepsilon}_{\text{eff}}^p = \sqrt{(4/3)\,D_2{}^p}$, and $f(D_2{}^p/D_0{}^2)$, obtained by inverting Eq. (3), is a relatively insensitive function of $D_2{}^p$ at strain rates low compared to D_0 which can be set to be from $10^4\ \text{sec}^{-1}$ to $10^8\ \text{sec}^{-1}$. Buckling is a relatively rapid process so it would be appropriate to ignore the thermal recovery terms in the hardening evolution equations for the determination of the applicable $E_T{}^p$. Multiplying Eq. (6) by $(dt/d\varepsilon_{\text{eff}}^p)$ and Eq. (7) by $u_{ij}(dt/d\varepsilon_{\text{eff}}^p)$, adding according to Eq. (5), and substituting into Eq. (9) leads to

$$E_T{}^p = [m_1(Z_1 - Z^I) + m_2(Z_3 - Z^D)]\ \sigma_{\text{eff}}\, f(D_2{}^p/D_0{}^2) \tag{10}$$

on the basis that the term involving $df(D_2{}^p/D_0{}^2)/d\varepsilon_{\text{eff}}^p$ can be ignored. That would be equivalent to calculating $E_T{}^p$ for strain rate magnitudes in the pre-buckling state so that the resultant value should be a lower bound since higher strain rates are realized at buckling. Again using the kinetic equation, Eq. (10) can be expressed as

$$E_T{}^p = [m_1(Z_1 - Z^I) + m_2(Z_3 - Z^D)]\ [\sigma_{\text{eff}}^2/(Z^I + Z^D)] \tag{11}$$

which is a function of σ_{eff}, Z^I and Z^D. In a given application, these values are obtained from the full set of equations including the recovery terms in the evolution equations for hardening. The material constants Z_1 and Z_3 are the saturation values for isotropic and directional hardening respectively; $E_T{}^p$ would decrease as hardening increases and tends toward saturation. It is noted that Eq. (11) is based on the plastic strain rate being constant with plastic strain and would not be applicable at the onset of inelastic deformation (the proportional limit) where the plastic strain rate is subject to large changes. In that region, the current plastic tangent modulus at a given overall strain rate could be obtained numerically from the full set of equations.

The instantaneous moduli L_{ijkl} controlling bifurcation (under increasing load in the sense of Shanley and generalized by Hill) for the small strain, incremental strain hardening J_2 flow theory depend on the elastic moduli, E, ν, the stress deviator J_2 (the effective stress), and the total tangent modulus E_T. This expression for L_{ijkl} is given, for example, by Eq. (3.46) of Hutchinson [5]. With E_T obtained from $E_T{}^p$ given by Eq. (11), the resulting bifurcation condition would then lead to a lower bound for buckling of the elastic-viscoplastic structure.

3 Application to creep buckling

It has been recognized for some time that creep deformations influence the geometry of a structure but have relatively small effect on the immediate response of the material to a load increment. Possible bifurcation during creep would therefore be controlled by the short time response characteristics which are distinguished from long time creep behavior. Comments to this effect have been made by a number of investigators such as Grigoliuk and Lipovtsev [6], Storåkers [7], Hoff [8], Obrecht [9] and Griffin [19], p. 341.

A procedure that has been used to examine possible bifurcation in the presence of creep, typified in the work of Bushnell [11], is to partition the inelastic strain into a time dependent creep component and a time independent plastic component. The creep component would influence the geometry of the structure over the time while a time independent effective stress-effective strain relation, presumed to be independent of the deformations, is taken to govern the short time response. The tangent modulus of that relation is the one that appears in the bifurcation criterion while creep deformations influence the state of stress through changes in the structural geometry.

An essential consideration in unified theories of viscoplasticity is that plasticity and creep are particular manifestations of time dependent inelastic behavior which can be represented by the same constitutive equations. Plasticity and creep are thereby inherently coupled in such theories so that it would be inappropriate to obtain the current tangent modulus that controls bifurcation from the overall response that includes creep deformations. Instead, the applicable "lower bound" tangent modulus should correspond to the rate of change of hardening with plastic strain at a steady plastic strain rate, i.e. be quasi-rate independent. For any constitutive theory for which the kinetic equation can expressed in the form,

$$\sigma_{\text{eff}} = Z \cdot f(D_2{}^p), \tag{12}$$

it follows that,

$$E_T{}^p = \frac{\dot{\sigma}_{\text{eff}}}{\dot{\varepsilon}_{\text{eff}}^p} = \frac{\dot{Z}}{\dot{\varepsilon}_{\text{eff}}^p} f(D_2{}^p) + \frac{Z}{\dot{\varepsilon}_{\text{eff}}^p} \frac{d}{dt}[f(D_2{}^p)], \tag{13}$$

and for $f(D_2{}^p)$ remaining constant,

$$E_T{}^p = \frac{\dot{Z}}{\dot{\varepsilon}_{\text{eff}}^p} \cdot \frac{\sigma_{\text{eff}}}{Z}. \tag{14}$$

The kinetic equation, (3), of the B-P model is of the form of Eq. (12), and Eq. (11), corresponding to Eq. (14), is the specific expression obtained for $E_T{}^p$ from that constitutive theory. As noted, that value would be a lower bound on the tangent modulus for buckling and depends only on the effective stress and the hardening variables.

In a typical creep buckling problem with constant load, the stress and the hardening variables would change with time due to alterations in the structural geometry and by plastic working of the material. Thermal recovery of hardening according to the full evolution equations (6) and (7) could also occur and has to be considered in the values for Z^I and Z^D. Following the procedure of Bushnell [11], the bifurcation criterion which contains $E_T{}^p$ according to Eq. (14), or specifically Eq. (11), needs to be tested at each increment of time to determine whether buckling has ocurred.

In the case of the ideal column, the stress and hardening variables would increase with loading. During primary creep at constant load (corresponding to constant stress for the perfect column), the hardening variables would continue to increase due to creep generated plastic work and the bifurcation condition could be reached at a certain time. At the onset of secondary creep, the hardening rate equals the rate of recovery so that Z and therefore the plastic strain rate remain constant for later times. If the bifurcation condition had not been reached at the onset of secondary creep, the proposed procedure would not admit the possibility of creep buckling of a perfect column. Then only an analysis based on initial geometrical imperfection would lead to creep buckling.

4 Application to buckling in modes generating non-proportional loading

The lengthy controversy over the relative applicability of deformation and incremental theories of plasticity has subsided somewhat with the recognition that detailed consideration of initial imperfections and well defined descriptions of the boundary conditions and the pre-buckling state enable the incremental theory to provide more realistic predictions, e.g. Gjelsvik and Lin [12], Tuğcu [13]. There are still shortcomings in the predictive capability of the standard incremental plasticity theory which are especially evident in problems for which the buckling modes generate significant non-proportional stress histories. A particular example is the initiation of stress components at buckling which were zero in the pre-buckled state.

A possible procedure to handle such conditions with yield surface theories is based on the assumption that the yield surface will develop large curvature (a "corner") at the loading point so that the direction of plasic straining would not be limited to the direction normal to the original yield surface. Such modified plasticity theories have become fairly sophisticated and appear to provide reasonable predictions to various buckling problems, e.g. Christoffersen and Hutchinson [14].

The rationale that a corner could develop at the loading point of a yield surface was partly motivated by certain physical theories of plastic deformation and implies that a kind of material instability or indeterminateness can occur under the circumstances. From a general viewpoint, the introduction of assumptions on changes of the shape of the yield surface is a search for special material effects to explain the deficiency of conventional incremental plasticity theory with a smooth yield surface and the associated normality condition to handle certain buckling problems.

Although experimental studies to examine the possibility of corners developing on yield surfaces have been inconclusive, other experiments on non-proportional loading have indicated definite non-conventional material effects upon a rapid change in straining direction in the inelastic range. Prime examples are the tests reported by Ohashi and Tokuda [15], Ohasi, Tokuda and Yamashita [16], Shiratori, Ikegami and Kaneko [17], Ohashi and Kawashima [18], and Ohashi and Ohno [19]. One of the interesting observations in these experiments, shown in Fig. 1 taken from Fig. 9 of [19], is that for a 90° rapid change in the direction of plastic straining in strain space at the same effective strain rate, the effective stress experiences a sudden drop to almost the level of the original proportional limit. That drop does not correspond to elastic unloading which would occur if the angular direction change were more than 90°. Also, there is a deviation from coaxiality between the directions of the plastic strain increment and the deviatoric stress which diminishes with continued straining in the new direction. After the rapid

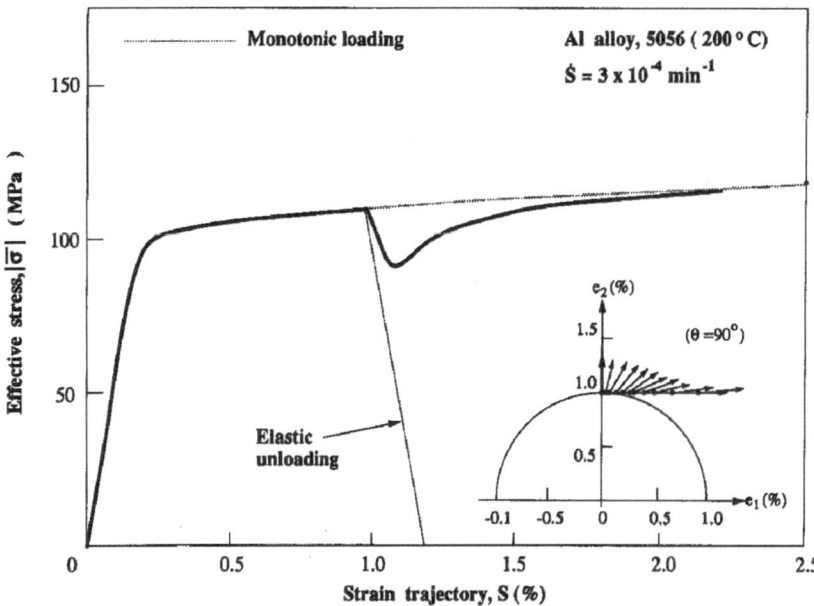

Fig. 1. Variation of the modified effective stress magnitude and direction along bilinear strain trajectory (from Fig. 9 of Ohashi and Ohno [19])

drop, the stress increases with continued straining at a rate higher than the slope of the effective stress-effective strain curve prior to the change in the stress direction. The effective stress recovers with continued straining along the second branch to what would have been its value without the interruption.

After the 90° change in direction of straining, constant strain conditions are maintained in the original direction which would induce stress relaxation. That appears to be accelerated by the generation of new slip directions and deactivation of the original ones leading to a drop in directional hardening and a corresponding drop in effective stress level. By this process, most of the directional hardening developed by the prior straining would be removed by a small increment of straining in the new direction. It was noted by Ohashi, Tokuda and Yamashita [16], p. 315, that the major part of the stress drop could be attributed to a reinforcing effect of the initiation of the second branch. Further straining along the second branch thereby encounters the isotropic hardening obtained by the initial deformation and additions to it and also the generation of directional hardening on the newly developed slip planes.

Sample numerical exercises have been performed using the basic form of the B-P equations, Table 1, for the case of a thin walled tube subjected to controlled extension and then suddenly to torsion at a total effective strain well into the plastic range, i.e. in the range from 0.5 to 1.0%. The results do show a rapid drop in directional hardening and effective stress at the change, but the amplitude was relatively small. There are various possible modifications that can be made to the equations to amplify the hardening drop such as considering the hardening saturation values to be functions of a factor that represents proportionality of the loading history. However, the detailed history of the rapid stress drop may not be required for the purpose of obtaining approximate values of the applicable hardening rates to use in the bifurcation criterion.

From the viewpoint of buckling, the tangent modulus to be used in the bifurcation criterion would depend on the controlling stress state. A bifurcation mode that involves only the pre-buckling stresses would be governed by the tangent modulus along the initial straining path, while that would not be appropriate if the mode develops non-proportional loading histories such as the generation of new stress components. In the latter case, the pre-buckling tangent modulus would not be physically relevant since it would not represent the stress state that governs bifurcation. Based on the preceding discussion of corner turning test results and on the Shanley/Hill concept of continued loading, bifurcation in such cases should be controlled by the hardening rate experienced in the new straining directions corresponding to the buckling mode. In this respect, the concepts leading to Eq. (10), (11) and (14) have to be altered in that the hardening rate in the immediate post-buckled configuration governs and not that in the pre-buckled state. Nevertheless, those equations could still be applied with the modified interpretation. From the experimental work cited and discussed, it appears that a good approximation for the applicable E_T^P can be obtained by setting $Z^D = 0$ in Eq. (11) for those bifurcation modes that involve abrupt changes in the relative proportions of the components of the stress increments. For cases in which all the stress components induced by buckling are smooth continuations of those in the pre-buckled state, the standard procedure of evaluating E_T^P should be applied using the values of Z^I and Z^D of the pre-bifurcation state. This means that problems that initiate the buckling mode stress components in the pre-buckled state by consideration of initial imperfections, pre-buckling deformations, exact boundary conditions, etc. are not influenced by the "corner effect" and can be treated in the direct manner.

A particularly interesting problem on which to exercise the suggested procedure is the torsional buckling of axially compressed columns of cruciform cross section. This problem has served for studies on the relative predictive capabilities of the incremental and deformation plasticity theories and on the role that small geometrical imperfections have on the results, e.g.

Onat and Drucker [20]. A series of tests on such columns of aluminium alloy 2024-T4 was reported by Gerard and Becker [21]. The graph of test results, Fig. 5 of that report, and the comparisons with the two plasticity theories is well known and reprinted in texts, e.g. Lubliner [22].

Detailed information on the material properties can be obtained in the paper by Gerard [23] in relation to similar tests, but detailed reportage of the cruciform column tests seems to be unavailable. A stress-strain curve for the specimen material and detailed numerical values of the tangent modulus as a function of stress and strain are given. The aluminium alloy was designated 24S-T in the 1946 paper [23] while the later (1957) and current identification is 2024-T4. The stress-strain curve, which is shown in Fig. 8 of the 1946 paper, [23], exhibits a smooth knee, i.e. transition from the elastic to the inelastic range, and the maximum axial strain for the cruciform column tests was about 1.3%.

Values for the full set of material constants to be used in the B-P constitutive theory for the aluminium alloy are lacking at the present time and will be obtained in a more thorough examination. Currently available are constants used by Aboudi [24] for 2024-T4 aluminium at room temperature that correspond to full isotropic hardening. These values are: $D_0 = 10^4 \text{ sec}^{-1}$, $n = 10, Z_0 = 340 \text{ MPa}, Z_1 = 435 \text{ MPa}, m_1$ (as defined in Table 1) $= (300/340) (\text{Mpa})^{-1}$. A value of the strain rate sensitivity parameter n of 10 corresponds to essential rate independence of plastic flow at low and moderate strain rates. As a consequence of the high n and accounting only for isotropic hardening, the knee of thee stress-strain curve deerived from the above constants is fairly sharp, Fig. 2 of [24], and does not adequately represent the smooth transition shown in Fig. 8 of [23].

Approximate values for the directional hardening constants could be obtained by assuming that part of the total hardening increment, $Z_1 - Z_0$, is directional. That fraction would be about 0.4 based on experience with similar materials. Also, experience indicates that the rate of directional hardening could be from 5 to 15 times that of isotropic hardening so that directional hardening dominates and saturates at the lower strain values, i.e. up to about $1 - 2\%$. Based on these considerations, a revised set of material constants for the preliminary calculations is: $D_0 = 10^4 \text{ sec}^{-1}$, $n = 10$, $Z_0 = 340 \text{ MPa}$, $Z_1 = 400 \text{ MPa}$, $Z_3 = 35 \text{ MPa}$, $m_1 = 1.0 (\text{Mpa})^{-1}$, $m_2 = 10.0 (\text{MPa})^{-1}$. With these constants, the stress-strain relation would exhibit a smoother transition from elastic to inelastic response in better agreement with the test indications.

As discussed previously, Eq. (11) is not accurate at the onset of significant inelasic deformation so the exercise should be carried out at the higher strain levels of the test data. There are some test points corresponding to the 1% pre-buckling strain level which can serve as the reference for the exercise. According to Table 2 of Gerard [23], at 1% strain: $\sigma = 50\,200$ psi (346 MPa), $E_T/E = 0.1$, and $E_S/E = 0.47$ where E_S is the secant modulus. From the equation relating $E_T{}^p$, E_T, and E,

$$(E_T{}^p/E) = (E_T/E)/[1 - (E_T/E)], \tag{15}$$

which ignores non-proportional effects, the corresponding $E_T{}^p/E$ is 0.11.

On the basis that Z^D saturates at the strain of 1%, i.e. $Z^D = Z_3$ prior to buckling, and the isotropic hardening does not increase much above Z_0 at that strain, a value for $E_T{}^p$ prior to buckling can be calculated from Eq. (11). At buckling into the torsional mode, Z^D developed by the previous straining will drop and is presumed to drop to zero. Placing $Z^D = 0$ into Eq. (11), with the other constants being the same, leads to an increase in $E_T{}^p$ by a factor of 7.5 over the pre-buckling value, or $E_T{}^p/E = 0.83$. From Eq. (15), the corresponding E_T/E would be 0.45 or 4.5 times the value of E_T/E prior to buckling. Such large increases of hardening rate are actually observed in corner turning tests, e.g. Fig. 9 of [19], Fig. 1 here. For the associated test condition, the operational torsional stiffness for buckling, which depends on the tangent shear modulus, is apparently less than the bending stiffness (for column buckling) at the same stress level despite the higher hardening rate due to the change in straining direction.

The derived increased hardening rate upon change in stress direction, $E_T/E = 0.45$ agrees well with the secant modulus value $E_S/E = 0.47$, and with test results on cruciform column buckling in torsion at that level of axial stress, Fig. 5 of [21], also, Fig. 5.3.4 of [22]. Similar calculations for other stress levels above the knee of the stress-strain curve also gave results comparable to corresponding E_S/E and tests. Those test results are plotted as a percentage of the torsional elastic buckling condition which is proportional to the elastic shear modulus G. The shear moduli in the inelastic range are the multiplying factors in the buckling criterion. When non-dimensionalized by G, they correspond to the moduli ratios obtained above. Of course, the drop in Z^D upon change of direction of straining at buckling should be obtained formally from the constitutive theory and not by the *ad hoc* method used in the numerical exercise. That was intended to show that suitable buckling values can be obtained by consideration of those material effects.

A tentative conclusion of this exercise is that proper treatment of realistic material behavior upon an abrupt change of straining direction may explain some of the apparent paradoxes for buckling of structures in the inelastic range. In particular, the tangent modulus to the effective stress-effective strain curve is considered to be the response characteristic that governs bifurcation in the inelastic range, where the modulus for the controlling stress state can be sensitive to the straining history at buckling. It should be possible to include those strain history effects in an elastic-viscoplastic constitutive theory.

Acknowledgement

The author would like to thank his colleague, Prof. Miles Rubin, for many useful discussions in the preparation of this paper.

References

[1] Tvergaard, V.: Plasticity and creep at finite strains. In: Proc. 17th Congress of Theoretical and Applied Mechanics, 1988 (P. Germain, M. Piau and D. Caillerie, eds), pp. 349–368, Amsterdam: Elsevier 1989.

[2] Tvergaard, V.: Rate sensitivity in elastic-plastic panel buckling. In: Aspects of the Analysis of Plate Structures (D. J. Sawe et al., eds.), pp. 293–308, Oxford: Clarendon Press 1985.

[3] Bodner, S. R., Naveh, M., Merzer, A. M.: Deformation and buckling of axisymmetric viscoplastic shells under thermomechanical loading. Int. J. Solids Struct. 27, 1915–1924 (1991).

[4] Bodner, S. R.: Review of a unified elastic-viscoplastic theory. In: Unified Constitutive Equations for Creep and Plasticity (A. K. Miller, ed.), pp. 273–301, London: Elsevier 1987.

[5] Hutchinson, J. W.: Plastic buckling. In: Advances in Applied Mechanics, vol. 14 (C. S. Yih, ed.), pp. 67–144, New York: Academic Press 1974.

[6] Grigoliuk, E. I., Lipovtsev, Y. V.: On the creep buckling of shells. Int. J. Solids Structures 5, 155–173 (1969).

[7] Storåkers, B.: Bifurcation and instability modes in thick-walled viscoplastic pressure vessels. In: Proc. IUTAM Symposium Creep in Structures, 1970 (J. Hult, ed.), pp. 333–344, Berlin, Heidelberg, New York: Springer 1972.

[8] Hoff, N. J.: The effect of geometric nonlinearities on the creep buckling time of axially compressed circular cylindrical shells. ASME J. Appl. Mech. 42, 225–226 (1975).

[9] Obrecht, H.: Creep buckling and postbuckling of circular cylindrical shells under axial compression. Int. J. Solids Struct. **13**, 337−355 (1977).

[10] Griffin, D. S.: Design limits for creep buckling of structural components. In: Proc IUTAM Symposium Creep in Structures, 1980 (A. R. S. Ponter and D. R. Hayhurst, eds.), pp. 331−345, Berlin, Heidelberg, New York: Springer 1981.

[11] Bushnell, D.: Bifurcation buckling of shells of revolution including large deflections, plasticity and creep. Int. J. Solids Struct. **10**, 1287−1305 (1974).

[12] Gjelsvik, A., Lin, G. S.: J. Engng. Mech. **113**, 953−964 (1987).

[13] Tuğcu, P.: Plate buckling in the plastic range. Int. J. Mech. Sci. **33**, 1−11 (1991).

[14] Christoffersen, J., Hutchinson, J. W.: A class of phenomenological corner theories of plasticity. J. Mech. Phys. Solids **27**, 465−487 (1979).

[15] Ohashi, Y., Tokuda, M.: Precise measurements of plastic behavior of mild steel tubular specimens subjected to combined torsion and axial force. J. Mech. Phys. Solids **21**, 241−261 (1973).

[16] Ohashi, Y., Tokuda, M., Yamashita, H.: Effect of third invariant of stress deviator on plastic deformation of mild steel. J. Mech. Phys. Solids **23**, 295−323 (1975).

[17] Shiratori, E., Ikegami, K., Kaneko, K.: Stress and plastic strain increment after corners on strain paths. J. Mech. Phys. Solids **23**, 325−334 (1975).

[18] Ohashi, Y., Kawashima, K.: Plastic deformation of aluminium alloy under abruptly changing loading or strain paths. J. Mech. Phys. Solids **25**, 409−421 (1977).

[19] Ohashi, Y., Ohno, N.: Inelastic stress responses of an aluminium alloy in non-proportional deformations at elevated temperature. J. Mech. Phys. Solids **30**, 287−304 (1982).

[20] Onat, E. T., Drucker, D. C.: Inelastic instability and incremental theories of plasticity. J. Aero. Sci. **20**, 181−186 (1953).

[21] Gerard, G., Becker, H.: Handbook of structural stability, part I − buckling of flat plates, NACA TN 3781, National Advisory Committee for Aeronautics, Washington DC, July 1957.

[22] Lubliner, J.: Mechanics of Solids, p. 313, New York: Macmillian 1990.

[23] Gerard, G.: Secant modulus method for determining plate instability above the proportional limit. J. Aero. Sci. **13**, 38−44, 48 (1946).

[24] Aboudi, J.: The effective thermomechanical behavior of inelastic fiber-reinforced materials. Int. Engng. Sci. **23**, 773−787 (1985).

Author's address: Prof. S. R. Bodner, Faculty of Mechanical Engineering, Technion-Israel Institute of Technology, 32 000 Haifa, Israel.

New Titles in Engineering

Hans Troger, Alois Steindl

Nonlinear Stability and Bifurcation Theory

An Introduction for Engineers and Applied Scientists

1991. With 141 figures. XI, 407 pages.
Soft cover DM 138,-, öS 966,-
ISBN 3-211-82292-5

The book tries to make the tremendous progress in the mathematical treatment of nonlinear dynamical systems in the field of stability theory available to scientists and engineers. A unified and systematic treatment of the different types of loss of stability of equilibrium positions of statical and dynamical systems and of periodic solutions of dynamical systems is given by means of the methods of bifurcation and singularity theory.

Alexander Weinmann

Uncertain Models and Robust Control

1991. 123 figures. V, 722 pages.
Cloth with slipcase DM 198,-, öS 1386,-
ISBN 3-211-82299-2

This monograph is devoted to plants and their approximate models and to the discussion of their uncertainties. The thrust of the book is on systematic representation of methods for robust control.

Franz Ziegler

Mechanics of Solids and Fluids

1991. 352 figures. XXI, 735 pages.
Cloth DM 148,-, öS 1040,-
ISBN 3-211-97529-2

This book offers a unified presentation of the concepts and most of the practical principles common to all branches of solid and fluid mechanics. A profound knowledge of applied mechanics as understood in this book may help to meet the needs for necessary versatility which the engineering community has to face in the modern world of high-technology

Prices are subject to change without notice

Springer-Verlag Wien New York

O.E. Barndorff-Nielsen, B.B. Willetts (eds.)

Aeolian Grain Transport

Volume 1: Mechanics

(Acta Mechanica/Supplementum 1)

1991. 79 figures. IX, 181 pages.
Soft cover DM 220,-, öS 1540,-
Reduced price for subscribers to "Acta Mechanica":
Soft cover DM 198,-, öS 1386,-
ISBN 3-211-82269-0

Volume 2: The Erosional Environment

(Acta Mechanica/Supplementum 2)

1991. 104 figures. IX, 181 pages.
Soft cover DM 220,-, öS 1540,-
Reduced price for subscribers to "Acta Mechanica":
Soft cover DM 198,-, öS 1386,-
ISBN 3-211-82274-4

These two volumes contain a collection of selected papers on a wide range of aspects of natural transport and re-sorting of sand and soil. Apart from one paper on avalanche and a small number in which sea waves are also an agent, the papers all deal exclusively with wind erosion.

One paper reviews in depth the contribution to the literature of aeolian transport which have been published during the last five years. The purpose of the workshop from which these volumes derive was to intimate and discuss work done during these five years but not yet published. The collection of papers was reviewed in light of these discussions and can therefore be taken to be a particularly well informed statement about the subject as it stood in mid 1990.

The five years 1985 - 1990 were a most exciting period in the study of wind erosion. Computational models were built of the whole process of bed-load transport, including saltation, based upon the basic features of grain aerodynamics and collision. This is the first successful portrait in such basic terms of the self-stabilising nature of any erosion process. Moreover, in the same period, an elegant and powerful account of the statistics of natural grain size distributions was applied to grain transport studies, with conspicuous success. Several important field problem classes were examined and several new observational techniques.

This activity, together with widespread concern about soil degradation and desertification in wind transported soils, will probably attract many new workers to the field during the next decade. This summary of its present status will be extremely helpful to them.

Prices are subject to change without notice

Springer-Verlag Wien New York